A
Pocket
History
of
Human
Evolution

HOW WE BECAME SAPIENS

SILVANA CONDEMI
FRANÇOIS SAVATIER

THE EXPERIMENT
NEW YORK

A Pocket History of Human Evolution: *How We Became Sapiens*
Copyright © Flammarion, Paris, 2018, 2019
Translation copyright © 2019 by The Experiment

Originally published in France as *Dernières Nouvelles de Sapiens* by Flammarion in 2018.
First published in North America in revised form by The Experiment, LLC, in 2019.

The Experiment, LLC | 220 East 23rd Street, Suite 600, New York, NY 10010-4658
theexperimentpublishing.com

THE EXPERIMENT and its colophon are registered trademarks of The Experiment, LLC. Many of the designations used by manufacturers and sellers to distinguish their products are claimed as trademarks. Where those designations appear in this book and The Experiment was aware of a trademark claim, the designations have been capitalized.

The Experiment's books are available at special discounts when purchased in bulk for premiums and sales promotions as well as for fund-raising or educational use. For details, contact us at info@theexperimentpublishing.com.

Library of Congress Cataloging-in-Publication Data

Names: Condemi, Silvana, author. | Savatier, François, 1961- author.
Title: A pocket history of human evolution : how we became sapiens /
  Silvana Condemi and François Savatier.
Other titles: Dernières nouvelles de sapiens. English
Description: New York : The Experiment, 2019. | Originally published in
  France as Dernières Nouvelles de Sapiens by Flammarion in 2018. |
  Includes bibliographical references and index.
Identifiers: LCCN 2019030095 (print) | LCCN 2019030096 (ebook) | ISBN
  9781615196043 (trade paperback) | ISBN 9781615196050 (ebook)
Subjects: LCSH: Human evolution--History. | Human beings--History.
Classification: LCC GN281 .C586413 2019 (print) | LCC GN281 (ebook) | DDC
  599.93/809--dc23
LC record available at https://lccn.loc.gov/2019030095
LC ebook record available at https://lccn.loc.gov/2019030096

ISBN 978-1-61519-604-3
Ebook ISBN 978-1-61519-605-0

Cover and text design by Beth Bugler
Cover illustration courtesy 3DMI/Shutterstock.com
Translation by Emma Ramadan
Illustrations by Thomas Haessig
Author photographs by Pauline Alioua-Flammarion (Silvana Condemi) and Ingrid
  Leroy (François Savatier)

Manufactured in the United States of America

First printing November 2019
10 9 8 7 6 5 4 3 2 1

Contents

Introduction   *v*

CHAPTER 1   A Biped Descends from an Ape   1
CHAPTER 2   Culture, the Evolutionary Accelerator   17
CHAPTER 3   My Big Head (Almost) Killed Me   27
CHAPTER 4   What Obligate Bipedalism Has Made Us   39
CHAPTER 5   Hunting Arouses All of the Senses   47
CHAPTER 6   The First Conquest of the Planet   59
CHAPTER 7   And *Homo sapiens* Emerged . . .   69
CHAPTER 8   The Spread of *Homo sapiens*
            Over the Entire Planet   83
CHAPTER 9   The Emergence of the Tribe   99
CHAPTER 10 War and the State   115

Conclusion   *129*
References   *133*
Acknowledgments   *143*
Index   *145*
About the Authors   *154*

Introduction

*H*omo sapiens is a strange animal. Our ancestors first lived in trees, then came down to explore the ground. Then they became bipeds and eventually explored the world—from there, the possibilities were endless. This changing behavior is one of the greatest enigmas there is, but we are in the process of unraveling it, with the help of amazing recent advancements in prehistoric sciences.

Through the extraction and sequencing of fossilized DNA we've learned that 40,000 years ago we were still sharing Earth with at least three other human species, and we know that Sapiens, an African species, crossbred with two other species outside of Africa. From new fossils we've also been able to prove that our ancestors did not originate exclusively in east Africa but were actually a Pan-African species. We've also discovered that Sapiens actually left its Pan-African cradle for the first time 100,000 years earlier than we had previously thought.

Though we have discovered so much about *Homo sapiens* over the years, there are still many unanswered questions about what caused us to become uniquely human. Was it climate change that propelled our supposed arboreal ancestors from the forests onto the ground in the savannas, setting in motion a series of complex anatomical changes? Was it bipedalism, which freed up our hands for other tasks? The use of tools? Our large brains? Did we become human because we were capable of empathy and cooperation?

For a long time, there were contradicting theories. Then, in 2015, we discovered something astonishing: 3.3 million years ago, in what would become present-day Kenya, stone tools were being made by hand. The oldest human fossil is estimated to be from 2.8 million years ago, so the hands that made these tools could not have been human. They could very well have belonged to *Australopithecus*, a prehuman. So it wasn't tools that made us human after all.

This news has led us to take a closer look at our ancestors, to discover new developments about Sapiens. In this book, we will focus on the progressive stages of hominization, the evolution-ary transformation of prehuman forms into *Homo* (including the cultural aspect of becoming human), which began in Africa more than 3 million years ago with *Australopithecus*. This was an astonishing transformation, one that gave rise to a strange and singular erect creature with strong cognitive abilities, whose most evolved form, Sapiens, carries the heritage of all its ancestors.

We all know that the primary role of Sapiens's advanced cog-nition is to help us survive. But where? In nature or in society? Sapiens is weak when alone in nature, but in groups we become the greatest predator that has ever existed. This seems ecological-ly impossible: a ubiquitous species that has transformed nature into its home—a home that has now reached global dimensions. In this book, we will explain this enigmatic evolutionary saga. This is the history of a cultural animal: you.

CHAPTER 1

A Biped Descends from an Ape

It was the increasing use of natural resources that pushed ancient primates toward the first human form. This not only led our ancestors toward bipedalism, walking upright on the ground, a more efficient way of moving, but it also triggered a self-reinforcing cycle: the more bipedal they were, the more success they had in collecting resources on the ground, which reinforced bipedalism, and so on. This alone, however, does not explain why humans became permanent bipeds.

In 1748, humans officially became animals for the first time. In his book *Systema Naturae* (*The System of Nature*), the botanist and zoologist Carl Linnaeus (1707–78) placed us in a group of related animal species—a genus—that he called *Homo*, and classified us as *sapiens*, meaning "wise." Today, *Homo sapiens* is the only existing human form.

As mammals—warm-blooded animals that nurse their young—Sapiens are members of the order of primates: apes with five fingers, front-facing eyes, and an upright torso when sitting. We don't know when primates first appeared, but we do know that they already existed during the Eocene era, which occurred from 56 to 33.9 million years ago. Where did they come from? We don't know this either, but

70 million years ago, when dinosaurs dominated Earth, there also existed a proto-primate called *Purgatorius*, a small animal the size of a mouse. So once the dinosaurs disappeared, modern mammals, including primates, were able to multiply their numbers.

Today, the majority of primates live in the tropics and are adapted to arboreal life, which suggests that humans' most ancient ancestors—hominid apes—lived in tropical forests where trees were tall and fruit was abundant. Most of today's hominid apes live in Africa, which points to an African origin for the human species.

Prehuman Hominins

There is overwhelming evidence, from the number of fossils of prehistoric hominins found in Africa, that suggests an African origin for *Homo*.

Today, the hominid family includes humans, bonobos, chimpanzees, gorillas, and orangutans (Figure 1). In addition, we have discovered many fossils, mostly of *Ardipithecus*, *Paranthropus*, and *Australopithecus*, which are our prehuman ancestors, namely hominins that are more closely related to us than they are to chimpanzees. So we can define the hominid family as all of the great apes that have a humanlike form and the same ability to walk on two legs.

There is evidence in the form of rare fossilized skeletal fragments that the hominin family already existed around 7 million years ago. What do all these fragments reveal? Actually, something quite fascinating: that the hominin evolutionary tree actually looks more like a bush, which shows that

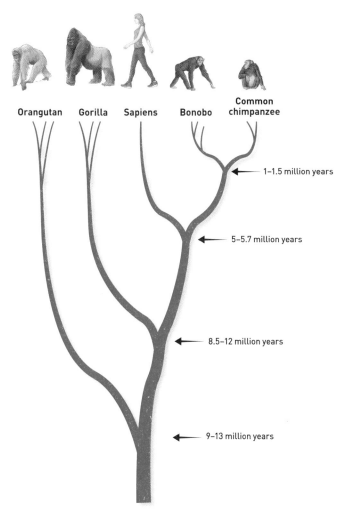

Figure 1

The hominid family tree

The various branches of hominids diverged millions of years apart, but we don't know exactly when, because we lack revealing fossils and the genetic data is not precise enough. The last common ancestor between *Pan* and *Homo*, for instance, lived between 7 and 5.5 million years ago.

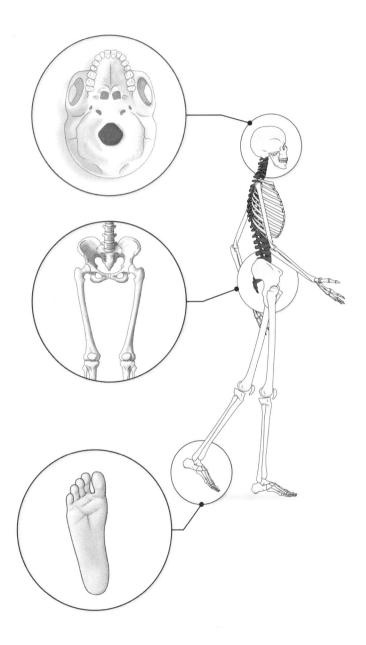

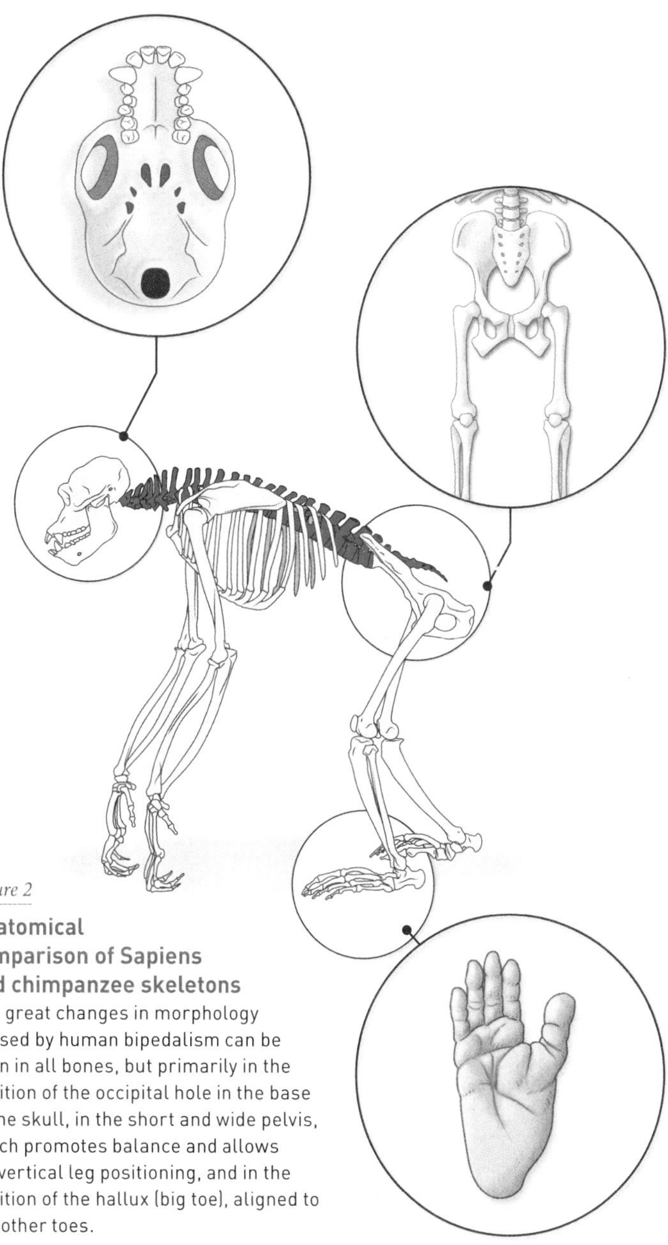

Figure 2

Anatomical comparison of Sapiens and chimpanzee skeletons

The great changes in morphology caused by human bipedalism can be seen in all bones, but primarily in the position of the occipital hole in the base of the skull, in the short and wide pelvis, which promotes balance and allows for vertical leg positioning, and in the position of the hallux (big toe), aligned to the other toes.

hominins passed through a series of significant evolutionary stages, periods during which several related species possessing nearly the same body type and lifestyles coexisted.

The first of these evolutionary stages is the transition from an arboreal quadruped posture to an imperfect bipedalism, meaning that these hominins were capable of walking on two legs but continued to be arboreal. Some specimens you might be familiar with are Toumaï, a *Sahelanthropus tchadensis* specimen from 7 million years ago, *Orrorin tugenensis*, from 6 to 5.7 million years ago, and *Ardipithecus*, from almost 5 million years ago. Of the oldest contender for earliest biped, Toumaï, we have only a fragment of femur bone and a skull, deformed by the sedimentary deposits not far from ancient Lake Chad. According to its discoverer, Michel Brunet, a paleontologist from the Collège de France, the fairly central position, under the skull rather than on the back, of its occipital foramen—the hole at the base of the skull where the spinal cord connects to the brain—strongly suggests that Toumaï had already adapted to erect posture and bipedalism on the ground (Figure 2). For this reason, Brunet sees Toumaï as one of the members of the hominin lineage, still closely related to the common ancestor we share with chimpanzees, but there is still some debate about whether this is adequate proof of bipedalism. It is hard to determine, since, for the moment, no other specimens of this species have been found.

Another ancient member of the human lineage is *Orrorin tugenensis*. Discovered by the paleontologists Brigitte Senut and Martin Pickford of the French National Museum of Natural History, *Orrorin* is known from a dozen fossil fragments, belonging to four individuals found in three locations in Kenya. With this hominin, it's the shape of the femur, which more closely resembles that of *Homo sapiens* than earlier hominid species and suggests an erect posture, while the ape-like curved thumb adapted to climbing trees also indicates an arboreal lifestyle. The *Ardipithecus* specimens unearthed in Ethiopia—*Ardipithecus kadabba* (5.8 million to 5.2 million years ago) and its likely successor *Ardipithecus ramidus* (4.4 million years ago), are much better preserved and much more informative (Figure 3).

It seems that the hominins that reached this stage of evolution were capable of only imperfect bipedalism: their feet were well adapted to walking but still had an opposable thumb like the foot of a chimpanzee. While this "walking thumb" severely limited *Ardipithecus*'s effectiveness while walking on the ground, it allowed them to climb trees quickly, as their hands still had the long fingers of a climbing monkey. One species of *Ardipithecus*, possibly *Ardipithecus ramidus*, then evolved into *Australopithecus*, most likely *Australopithecus anamensis*, around 4.5 million years ago (Figure 3).

In the Footsteps of *Australopithecus*

Fittingly, the second significant stage of hominin evolution is the first true biped, *Australopithecus*. While its genus name—*Australopithecus*, meaning "ape of the south"—is a reference to its first discovery in South Africa in 1924, *Australopithecus* specimens have also been found in the Great Rift Valley in East Africa (Figure 4). As crazy as it might seem, the oldest evidence we have of *Australopithecus* is not a fossil at all

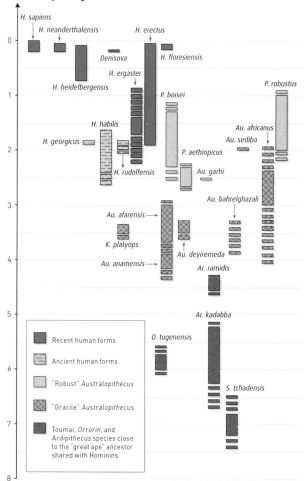

Millions of years ago

H. sapiens
H. neanderthalensis
H. erectus
Denisova
H. floresiensis
H. heidelbergensis
H. ergaster
P. boisei
P. robustus
H. habilis
Au. africanus
Au. sediba
H. georgicus
P. aethiopicus
H. rudolfensis
Au. garhi
Au. bahrelghazali
Au. afarensis →
K. platyops
Au. deyiremeda
Au. anamensis →
Ar. ramidis
Ar. kadabba
O. tugenensis
S. tchadensis

Recent human forms

Ancient human forms

"Robust" *Australopithecus*

"Gracile" *Australopithecus*

Toumaï, *Orrorin*, and *Ardipithecus* species close to the "great ape" ancestor shared with Hominins

Figure 3

The human and prehuman species over time

This hominin family tree includes all *Homo* species and all their early relatives including *Australopithecus*, *Paranthropithecus*, *Ardipithecus*, *Orrorin* and *Sahelanthropus*. Today the only living *Homo* species is *Homo sapiens*, but genetic studies have shown that Sapiens interbred with Neanderthals and Denisovans.

but a few footprints that have been preserved: they belong to three specimens of *Australopithecus afarensis*, the same species as the famous Lucy. Around 3.8 million years ago, at a site called Laetoli in what is now Tanzania, the Sadiman volcano covered the earth in a six-inch-thick bed of ash, which preserved the footprints of three *Australopithecus* specimens who walked through it together.

These footprints are especially fascinating because they resemble those of modern-day humans (Figure 5). The first toe on the foot—the hallux—is not opposable as it is in great monkeys, but instead lines up along the four other toes, like that of Sapiens. Therefore, while the two plantar arches (longitudinal and transverse) are not as pronounced, the *Australopithecus* foot is very similar to ours. We can also see from these footprints that the point of impact on the ground was made toward the heel, which implies that *Australopithecus* had not yet adopted the characteristic walk of the human: toes first, followed by the tightening of the plantar arch and then the heel on the ground.

Their hands also resemble ours, except that the first phalange of the thumb doesn't allow for as much movement as that of a human hand, and the fingers are long and slightly curved. This hand, along with long arms, narrow shoulders, and an upwardly narrowing thorax, shows an agility suited to climbing trees.

These characteristics are found in other species of *Australopithecus* (Figure 4), most notably in those originating in South Africa, *Australopithecus africanus* (3 to 2.6 million years ago) and *Australopithecus sediba* (2 million years ago),

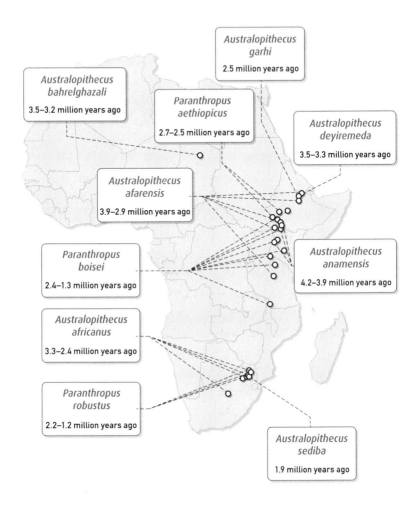

Figure 4

Main sites of *Australopithecus* "Gracile" and *Paranthropus* (once named *Australopithecus* "Robust") in Africa
No *Australopithecus* sites have been found in north and in western Africa. It is in one of these *Australopithecus* species, probably *Australopithecus anamensis*, that the process of hominization occurred.

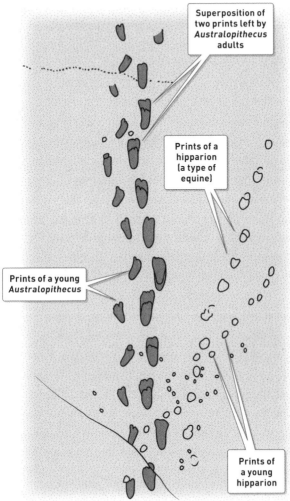

Superposition of two prints left by *Australopithecus* adults

Prints of a hipparion (a type of equine)

Prints of a young *Australopithecus*

Prints of a young hipparion

Figure 5

***Australopithecus* footsteps in a bed of ash in Laetoli**

These footsteps, of three *Australopithecus* and two hipparion, were found 30 miles (45 kilometers) south of the Olduvaï Gorge in Tanzania by archeologist Mary Leakey in 1974. Their age of 3.7 million years was determined by uranium-argon dating.

but with the latter, fairly long lower limbs indicate a substantial body size, which will become a crucial characteristic of fossils connected to the *Homo* genus.

Even though *Australopithecus* had hands and feet comparable to ours, it doesn't seem that they were exclusively bipedal. Some paleontologists have compared them to bonobos, meaning that they would have had a highly developed social life and spent most of their time on the ground foraging but remained largely dependent on forest trees. In particular, they would have found safety in the heights. Which makes us wonder: did they build nests in the trees to sleep in, like chimpanzees? It is with this chimpanzee behavior in mind that, after reexamining Lucy's bones, John Kappelman and his colleagues from the University of Texas at Austin suggested, in a 2016 *Nature* article, that she died by falling from a tree.

What triggered the evolution toward bipedalism among certain hominid populations? Yves Coppens of the Collège de France, who discovered, along with the American anthropologist Donald Johanson, the famous *Australopithecus* Lucy, suggested that the cause is climate change, brought on by the formation of the Great Rift Valley. According to a theory that he calls "East Side Story," the consequence of this geological event was a partial replacement of the forest by the savanna in East Africa. As a result, the quadrupedal primates accustomed to life in the trees had to move into open spaces, which was easier on two legs, setting in motion their evolution toward bipedalism.

But Coppens's theory, also called the *savanna hypothesis*, is not in agreement with the most recent environmental data, which shows that Orrorin, *Ardipithecus*, and even Lucy occupied a mosaic territory, with dense woodland, bush land, and savanna close to bodies of water. Some researchers have suggested that our upright posture began in our past arboreal life and was an advantage for standing on trees while feeding. Therefore, we don't yet have a definitive answer to this question.

Homo, the Obligate Biped

The third significant evolutionary stage—one that we still haven't left behind—is that of the obligate biped, fully committed and adapted for walking only on the ground with two legs. Much more complex than it seems, biomechanically speaking, this method of movement is absolutely unique to terrestrial vertebrates. In fact, it is the defining characteristic of the *Homo* genus and distinguishes us from all other hominids. But how did it come about? We think that it was through a gradual refinement of *Australopithecus*'s imperfect bipedalism. This opinion is backed up by an intense study of *Australopithecus afarensis* foot bones carried out by Carol Ward of the University of Missouri. She suggests that Lucy's relatives had arched feet, meaning they'd already adopted a way of moving that is similar to ours, and through this, became semi-human.

The shift from facultative to obligate bipedalism is evident in most ancient human species, notably in *Homo habili*s (2.8 to 1.44 million years ago), who lived in southeast Africa. While its cranial capacity was modest, closer to that of *Australopithecus* (a third of ours), its feet were very similar to a present-day human's. According to the nearly complete fossils that are available to us, the *Homo habilis* foot was rigid, possessing a vaulted plantar and bones similar in proportion to ours. This indicates obligate bipedalism, according to Yvette Deloison, a specialist in hominid locomotion from the French National Center for Scientific Research (CNRS).

Almost as old as *Homo habilis* and confined to East Africa, *Homo rudolfensis* (2.45 to 1.45 million years ago) was more robust and had a cranial capacity that was slightly more developed. It was also clearly an obligate biped. Thus, it's clear that obligate bipedalism is what distinguishes the *Homo* genus from *Australopithecus*. That's why the dominant hypothesis among paleontologists is that the evolution toward increased bipedalism marked the start of the transition from *Australopithecus* toward a human form.

Homo, Master of the Earth

Nevertheless, one question remains: what triggered this evolution? Because something pushed one lineage of *Australopithecus* to use natural resources on the ground to a greater extent and to expend less energy doing so. And lots of research supports this viewpoint. In 2010, a team led by David Raichlen of the University of Arizona concluded that in covering the same distance, a human expends only a quarter of the energy that a chimpanzee does on four legs.

These same researchers also found that chimpanzees expend the same amount of energy whether they walk on four legs or two. Thus, regardless of whether they take small steps while standing upright or use all of their muscles to run on four feet, they always expend more energy than humans. As soon as ancient hominins started to make use of ground natural resources, evolution could only move toward decreased energy expenditure, meaning increased bipedalism.

The need to traverse larger areas created a selective pressure toward increased bipedalism. In Kibale National Park in Uganda, a team led by Sabrina Krief of the French National Museum of Natural History has observed that the typical habitat of a chimpanzee covers about 8 square miles (20 square kilometers), but most of the time they use only a quarter of it. By comparison, ethnographic research has estimated that, depending on the climate (in warm areas, resources are abundant, but in colder climates, resources can be scarce), a group of hunter-gatherers would have needed to utilize up to 500 square miles (1,300 square kilometers). Bipedalism would also have been more advantageous for surveying an environment populated by a number of predators that would have been dangerous to fragile prehumans (Lucy was only three feet tall).

A Powerful Evolutionary Accelerator

The move toward a greater use of natural resources, however, does not entirely explain our evolution toward obligate bipedalism. Let's remember that during the second significant stage of evolution, several related species of *Australopithecus* lived simultaneously, and though they were already mostly bipedal, not all of them developed obligate bipedalism. The most convincing explanation for this curious phenomenon is that one lineage of *Australopithecus* "accelerated" their evolution, advancing more quickly than the others toward obligate bipedalism and therefore toward greater success at collecting natural resources on the ground. Over time (a span of about a million years), this lineage would have become so dominant that they would have first triggered the extinction of every related species in the same ecological niche, and eventually those outside it.

CHAPTER 2

Culture, the Evolutionary Accelerator

Culture caused the emergence of the human lineage by accelerating the evolution of certain Australopithecus *toolmakers. Their enhanced stature and increased cerebral volume triggered the process of hominization, which gave rise to larger and more cooperative human social groups, maintained by "linguistic grooming."*

What transformed certain lineages of *Australopithecus* into humans? We believe it's quite clear: culture. Culture refers to any set of behavioral traits, symbols, and ideas shared by an animal group across space (within members of the same group) and time (over generations). According to this definition, groups of dolphins or chimpanzees also have cultures, though it hasn't had the same evolutionary impact for these animals that it has had for humans. Could culture have developed differently for certain lineages of *Australopithecus* as bipedalism freed up the use of their hands? Yes, as we learned in 2015, when it was discovered that stone tools were being made before the known emergence of *Homo*. In a site called Lomekwi 3, not far from the western shore of Kenya's Lake Turkana, a team led by the prehistorian Sonia Harmand of CNRS and Stony Brook University discovered the oldest known stone tools, fashioned 3.3 million years ago, and therefore evidence for change in the hominin lifestyle. The oldest discovered fossil of the *Homo* genus is the partial jaw LD 350-1 still carrying six teeth, found in Ledi Geraru, Ethiopia, which dates back to 2.8 million years ago. The tools from Lomekwi 3—which indicate a stone culture now known as Lomekwian—are thus half a million years older than the oldest fossil evidence of the *Homo* genus. The simplest explanation for this confounding development is that these tools were carved by the only hominids around at that time, some form of *Australopithecus*.

Cultivated, but Flat-Faced

Sonia Harmand's team drew attention to the nearby discovery of *Kenyanthropus platyops*, the "flat-faced man from Kenya," a fossilized specimen that doesn't display the long face typical of apes but rather a much shorter face, in which some paleontologists see the characteristics of both *Australopithecus afarensis* and early *Homo*. *Kenyanthropus platyops* seems to be in the same evolutionary stage as *Australopithecus*, specifically *Australopithecus afarensis*, which lived in the same area at the same time. Since there couldn't have been aliens in Kenya 3.3 million years ago, the most logical conclusion is that either *Australopithecus afarensis* or *Kenyanthropus platyops* made the tools found at Lomekwi 3. But no matter which group made them, we can definitively say that the first toolmakers date back to the evolutionary stage of *Australopithecus*. Toolmaking is a cultural phenomenon, from a culture that preceded *Homo*.

Can we actually believe this? Yes, and here's why: since the primatologist and anthropologist Jane Goodall studied the chimpanzees of Gombe Stream National Park in Tanzania in the1960s, we've known that nonhuman hominids use tools. But they use tools of circumstance, such as clubs or rocks, for tasks like digging holes or cracking nuts open. The stones tools at Lomekwi 3, by contrast, were intentionally knapped stone tools to produce sharp edges. They were made using two techniques: direct impact on a stationary anvil (hitting the stone on a block to carve it) and bipolar impact on an anvil (shaping the stone with another while holding it against a block). This complexity implies that these techniques were mastered by the group and thus transmitted within the framework of a cultural tradition (Figure 6).

Moreover, these two methods of carving illustrate that, like the evolution of the species, technical evolution is also "bushy": the techniques diversify constantly, creating branches that die quickly, but the main trunk continues to grow. As the prehistorian Hélène Roche of the French National Center for Scientific Research (CNRS), who played a preeminent role in this research, explains: "Plio-Pleistocene behavioral evolution is a complex thing, as complex as biological evolution." The Plio-Pleistocene, we should note, is the era of Lomekwi 3.

Now, a stunning picture emerges: it's very likely that from the end of the Pliocene epoch (5.3 to 2.58 million years ago) to the beginning of the Pleistocene (2.58 million to 11,700 years ago), several species of *Australopithecus* fashioned stone tools and used them—if not for hunting, then at the

1 Freehand carving of a chopper

2 Freehand carving of a bifacial edge (successor to the chopper)

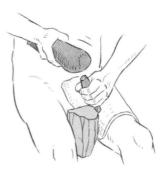

3 Blade shaping with a soft striker (antler)

Figure 6

Making stone tools

Stone knapping techniques were well advanced long before the first *Homo sapiens*. These techniques included direct percussion using free hand knapping first to sharpen a rock (1), then to carve a biface (2), and indirect percussion with a soft striker (3).

very least for cutting, harvesting root vegetables, and butchering dead animals.

The bones of ancient cattle, from 3.4 million years ago, which bear cut marks from stone tools, strongly indicate that this was the case. These were found at the Ethiopian site of Dikika, a site where a female specimen of *Australopithecus afarensis*, dating from around 3.3 million years ago, was found. Further evidence that *Australopithecus* made tools is the discovery, in the Bouri Formation near Ethiopia's Awash River, of *Australopithecus gahri*, which dates to 2.5 million years ago, along with some carved stone tools (though the geological association between the fossil and the tools is still in dispute). But it's clear that one of these toolmaking *Australopithecus* species must be the origin of the human stone technology that initiated Sapiens's technical domination of the world.

The Hominization of *Homo*

Armed with these archaeological arguments and observations, we can now suggest a broad description of the mechanism that gave rise to the human genus: hominization. *Homo* appeared from within the lineages of toolmaking *Australopithecus*, the result of the conjunction of bipedalism and cultural transmission. Because they could now walk upright, these individuals could make better use of their environment on a grand scale, leading to increased cooperation between members of their groups. This enabled increased imitative capacity (cognition), the development of manual skills (through cognition and the evolution of the hand, most notably to produce stone tools), and the ability to run upright (through the evolution of the body). These developments, in turn, increased cooperation in the group, leading to greater exploitation of the terrain. In brief, there began a strong cycle of reinforcement, resulting from selective social pressures. This is how the long evolution toward bipedalism led to prehuman forms and then ultimately humans. (Pre)human biology and (pre)human culture were coevolving even before *Homo*.

Bigger Bodies, Bigger Heads

It's a good theory, but is there any evidence of this reinforcing cycle in the fossil record? Certainly—if we subscribe to the idea that a larger stature would have been advantageous in the tool-assisted exploitation of ground natural resources. Greater body size, and therefore greater arm length, allows a human to throw a projectile more powerfully (the human shoulder enables humans to throw objects much faster than any other species), to strike more forcefully with a rock, to dig deeper into the soil with a stick, to handle longer spears, and so on.

If we follow this line of thinking, hominization occurred through the selection of certain prehuman lineages, then larger and larger humans, until a biological optimum was reached. Moreover, the larger size of these hominids made it easier for them to take advantage of a greater diversity of resources (meat, roots, honey, nuts, and fruit, for example). Stone tools, better organization, and coordination in the group helped them extract these resources from a larger area. This implies that they had an increased cognitive ability.

We know that there was a definite increase in human stature over time, from 4.3 feet (male *Homo habilis*) to around

5.6 feet (male *Homo sapiens*). This was accompanied by an increase in cerebral volume, from 400 cubic centimeters (male *Homo habilis*) to around 1,350 cubic centimeters (male *Homo sapiens*). Thus, cranial capacity continued to increase until it reached a biological optimum with the brain of Sapiens (the now-extinct Neanderthal species represented the maximum cranial capacity in *Homo*, close to 1,700 cubic centimeters).

To be clear, elephants have much bigger brains than ours, but they aren't any smarter. To assess the enhancement of cognitive capacities that accompanied enhanced cerebral volume, we must apply the encephalization coefficient, which establishes a relationship between the weight of the brain and the weight of the body.

In gorillas, this ratio is 1/230. It's between 1/90 and 1/180 in chimpanzees, but just 1/45 in today's humans. So, if we consider that Toumaï's evolutionary stage is comparable to a chimpanzee, then humans have multiplied their cranial capacity four times over 7 million years. This brain growth was even more accelerated over the last half million years, starting with *Homo heidelbergensis*, considered by most paleoanthropologists to be an ancestor common to both Neanderthal and Sapiens (Figure 7).

We see that successive species of (pre)humans offer evidence to support the existence of the cycle of reinforcement: increased bipedalism → larger accessible territory → greater size, cognition, and mobility to utilize resources → increased bipedalism, and the cycle repeats.

There is another indicator to support the existence of this cycle, revealing that a large proportion of primate cognition

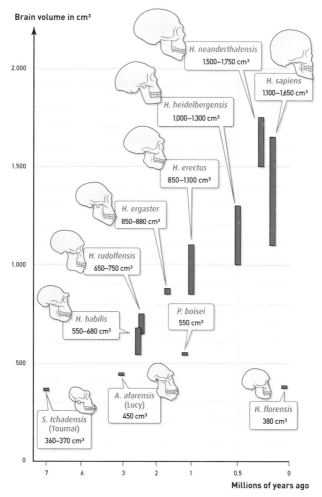

Figure 7

The evolution of cranial volume in hominins

A gradual increase in cranial volume has accompanied human evolution. From the common ancestors of *Homo sapiens* and Neanderthals on, around 500,000 years ago, the volume of the human brain stopped being proportional to the body's mass, growing bigger and bigger.

developed in service to the social organization and the social adaptation of the individual. In 1993, while at the University of London, professors Leslie Aiello and Robin Dunbar studied the relationship between the size of a primate's social group and the thickness of the neocortex, the outermost layer of the brain assumed to be the seat of cognition. To function, a group of primates needs to invest in mutual body care (delousing, for example), which establishes and maintains interpersonal connections. With chimpanzees, this social grooming takes up more than 16 percent of their time. Aiello and Dunbar established a mathematical law from the behavioral data of contemporary primates and applied it to hominins. They came to the conclusion that *Australopithecus* spent 20 percent of its time on social grooming; this time then increased to 45 percent for *Homo neanderthalensis* and *Homo sapiens*. However, we no longer spend half of our time delousing each other in order to maintain social stability, notably since our social groups are typically made of hundreds of people (in terms of relationships we maintain) or more (if we include the relationships we don't maintain).

According to the researchers, we replaced social grooming with the development of a new, more efficient type of bonding: language, a way to "symbolically groom" lots of people at once. Analysis of human conversations shows that about 60 percent of our time is spent gossiping about relationships and personal experiences. Aiello and Dunbar suggested that language evolved to allow individuals to learn about the behavioral characteristics of other group members more rapidly than is possible by direct observation alone.

CHAPTER 3

My Big Head (Almost) Killed Me

Culture modified our biology, which in turn evolved to enable more culture. This phenomenon can first be seen in our physiology, which requires the storage of fat for our large brain to function properly. It's also seen in our reproduction, which, with the increase in the size of the brain, has been pushed beyond the limits of primate obstetrics. Because the time it takes for children to reach maturity increased, child rearing began to require more and more cooperation, becoming a collaborative social activity.

In our long evolutionary history, it's important to be aware of a key point regarding hominization: culture deeply altered our biology from its primate origins. Selective pressures that began with the emergence of obligate bipedalism, and then hominization, were so intense that they reconfigured not only our head, our S-curved spine, our long legs, short arms, and narrow pelvis, but also our digestive system and our cognition. Our radical brain development resulted in profound physiological and behavioral remodeling.

Our evolution has pushed the development of our skull to its biological limits, beyond what seems physiologically possible, especially for female bodies and in terms of metabolism.

We all know that childbirth is painful and often dangerous for women. Labor lasts an average of 9.5 hours, which is five times longer than for gorillas, chimpanzees, or orangutans. This can be easily explained: the voluminous human brain requires a big head, which passes through the pelvic canal with difficulty. This has resulted in cranial bones that are soft and flexible in newborns, allowing some for deformation of the head, to facilitate childbirth. Once labor has begun, a human baby's head, which is slightly too large, must rotate to be able to descend the birth canal.

Childbirth would not be possible if development in utero were any longer, since if our babies were born at the same stage of development as our chimpanzee cousins, their heads would be too large to pass through the birth canal. As a result, our babies are born with a skull and brain that are still unfinished, almost absurdly immature compared to those of other animal species (Figure 8).

Once a baby is born, its brain continues to grow in size during the first seven years, while the little human is no longer isolated in the uterus but surrounded by relatives. So the human brain reaches full development while the child is already under the influence of social life. To perfect cerebral development, humanity has replaced the uterine womb with the "social womb." One peculiarity that partly explains our impressive cognitive development is the fact that this growth continues until our brains have around eighty-six billion neurons, compared to just six billion for our cousin the chimpanzee. A veritable thinking machine, the human neocortex represents 33 percent of the total volume of the brain

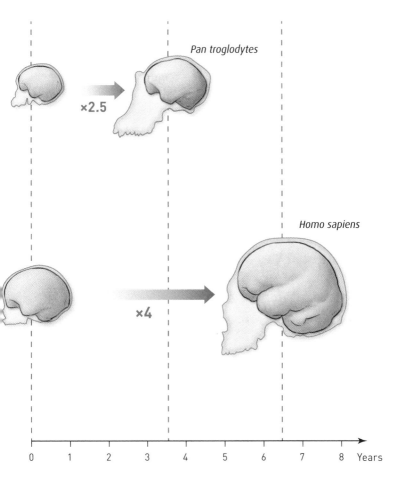

Figure 8

Growth of Sapiens and chimpanzee brains from birth to adulthood
The chimpanzee brain reaches its adult size in around 3.5 years. In
contrast, the Sapiens brain grows a lot during the first year, reaching its
adult size only around 6.5 years after birth, but its development remains
incomplete until 18 years after birth. Throughout life, the human brain has
the ability to change in reaction to stimuli.

in humans, compared to only 17 percent in chimpanzees. By the time a child is one year old, its brain is two-thirds of the size of an adult brain. The brain reaches its adult size between 1 and 6 years of age, but the region of the brain called the prefrontal cortex, which is responsible for the temporary storage of information, continues to mature. This process is mostly complete at adolescence but continues after with the acquisition of new learning. At the same time, once it has reached its final size, the brain continues to mature by "pruning": eliminating neural connections that are not regularly activated by stimulation. We know that our brains can suddenly and constantly reconfigure themselves at any age, according to our lived experiences.

Reproduction Boosted by Longevity . . .

From time immemorial, our big brain has cost women and babies their lives in childbirth, but that hasn't stopped us from reproducing. In fact, there are now almost 7.5 billion individuals on Earth. So how can we explain this paradox? The slow development of the human brain is only really possible if the parents live long lives. Therefore, humans practice what ecologists called a K-selection strategy, meaning that we have few offspring, long gestation periods, long-term parental care, and a long period before reaching sexual maturity. Species that follow the opposite strategy, r-selection, have many offspring, short gestation periods, less parental care, and a short period before reaching sexual maturity.

Unlike these r-strategist species, like certain fish that lay five hundred thousand eggs—only to have most of them eaten by predators—the human reproductive strategy combines a low number of births with strong parental investment. This is why, for millions of years, human culture has played a crucial role in raising children. In fact, our child rearing is collaborative: children are, at certain times, put in the care of other women, men, their siblings (whose education they benefit from most closely), and especially, if they're around, their grandmothers. Rachel Caspari of the University of Central Michigan and Sang-Hee Lee of the University of California, Riverside, have suggested that older individuals only became more common in human groups starting at the beginning of the Upper Paleolithic era (between 40,000 and 10,000 years before our current era). Increased longevity in humans means that older individuals began to play a larger role in child rearing, educating children based on their own experiences and therefore contributing on a grand scale to the evolution of human culture.

Grandmothers have helped raise their grandchildren for a long time. This investment has been so effective in terms of infant survival that natural selection in humans has led to a relatively early end in female fertility, long before death. This largely explains the human phenomenon of a very long menopause, which differs from that of other primates, such as apes, who generally die fairly soon after entering menopause.

. . . and Enhanced by Fat

Human lineages with fat women (and fatter men) were also favored by natural selection. In 2010, a team led by Richard Wrangham, a primatologist in the Department of Human

Evolutionary Biology at Harvard University, showed that sedentary humans are fatter than all other primates, even when those primates sit in cages at the zoo all day. According to Wrangham, this human trait—so despised today—is explained by the fact that, thanks to their ability to store energy in many different areas of their bodies, females, whose body fat is 25 percent greater than males during reproductive ages, are able to have more pregnancies in a row and therefore more babies than female apes, even though human sexual maturity is reached later and the gestation process takes longer. Once again, the large human brain is the reason for this increased fat storage: it creates a guarantee that, during famine or at other difficult times, the baby will develop appropriately and the brain of the mother will function well during breastfeeding, increasing the chances of survival for both mother and child. When conditions are good, a human baby is already fat at birth—which we instinctively find adorable.

Basically, the human brain needs considerable energy to function properly. Even at rest, the temperature-sensitive brain requires a certain amount of energy to keep it at body temperature. We also need energy to maintain our body's vital functions (breathing, temperature regulation, body maintenance) and vital organs (heart, lungs, kidneys, liver). This vital minimum energy—which we call basal metabolic rate—depends on the size of the individual, their age, their gender, and climatic conditions. Children have a basal metabolic rate that is much higher than that of adults, especially while the brain is maturing from infancy to age 7. Even though it represents just 2 or 3 percent of our body weight, the brain itself consumes 20 to 25 percent of our basal metabolic rate. Of course, these needs are even greater when the brain is hard at work.

This is huge in comparison to other primates. In 2016, a team led by Herman Pontzer of the City University of New York compared the energy expended by humans and anthropoid apes at rest and discovered that a human burns, on average, 400 more calories each day than a chimpanzee or bonobo, 635 more calories than a gorilla, and 820 more than an orangutan. This hypermetabolism (or increase in calories) can be explained by the energy needs of our sizable brain.

The First Organic Meat

All paleontologists agree that the introduction of animal proteins to the human diet played a crucial role in the evolution of our enormous thinking machine. In one single serving, meat provides not only energy in the form of protein and fat, but also many of the vitamins and minerals the body needs to function.

Archaeological evidence suggests that throughout history, hunter-gatherers sought to obtain as much high-energy meat (meat high in fat content) as possible. Despite the considerable risks involved, they were determined to slaughter large and often dangerous animals, including mammoths, cattle, rhinoceros, seals, and whales, becoming the world's top predator. One passing observation along these lines: when Sapiens began to domesticate animals for meat around 10,000 years ago, they first focused on animals high in fat content, such as pigs, cattle, sheep, and goats, before turning to birds and horses. Similarly, the first plants they domesticated were also energy-rich (wheat at first, followed by lentils and beans).

In a 1995 *Current Anthropology* article, Leslie Aiello and Peter Wheeler presented a theory known as the *expensive tissue hypothesis*, which states that the evolution of *Homo* re-directed the majority of available energy to the larger brain and away from the other organs, most notably the digestive system. Beginning with *Homo ergaster* (1.9 million years), the growth of the brain has occurred at the expense of rela-tive muscle power and intestinal length (Figure 9). The large intestines of apes are suitable for processing large amounts of leaves and ripe fruits, food that is relatively nutrient-poor. The gut of *Homo ergaster*, who had access to greater range of resources (including lake fish and shellfish), had already decreased in size compared to *Homo habilis* and *Australo-pithecus*. As humans became more carnivorous, the length of our intestines continued to decrease. And since we've already referenced the impressive slaughter of mammoths by hunt-ers, we should stress that we have always been able to have enough energy for our muscles for such hunting achieve-ments, even if we've become weaker than other great apes: the energy expended by muscles is just 20 percent of our basal metabolic rate but 40 percent of gorillas'.

Mastering Fire

In addition to procuring energy-rich foods, our ancestors learned to take better advantage of that energy, by taming fire. Cooking is a powerful way to prepare food, making it easier to chew and more digestible, as well as enhancing its caloric

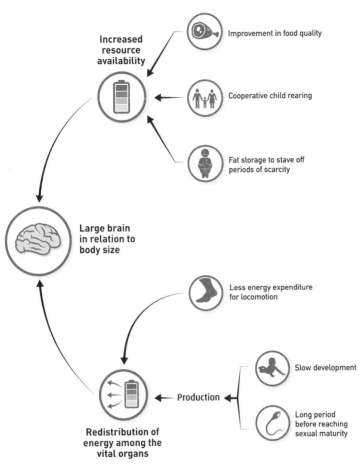

Figure 9

**Evolutionary and adaptative strategies to accommodate
our big brain**

The human brain needs considerable energy to function, around
20 percent of our total energy. Various biological and cultural changes that
occurred during the evolution of *Homo* made it possible to save and/or
redirect the energy available to the larger brain and away from of the other
organs, most notably the digestive system.

value. Studies show, for example, that 35 percent of cooked starch can be digested, compared to 12 percent of raw starch, and 78 percent of cooked protein can be digested, compared to 45 percent of raw protein. Thanks to cooking, today our digestive system requires energy that's barely equivalent to 10 percent of our basal metabolic rate.

To what point can we trace the taming of fire? In the Middle East, the oldest undeniable hearths—those at Gesher Benot Ya-aqov in Israel, for example—date back to some 790,000 years. In Europe, those in Přezletice in the Czech Republic date to 700,000 years ago; in Menez Dregan in Brittany and Vértesszöllös in Hungary to 450,000 years ago. In China, at Zhoukoudian, there's a hearth that dates back 420,000 years.

Consequently, in Europe, the shared ancestor of *Homo neanderthalensis* and *Homo sapiens*—*Homo heidelbergensis*—mastered fire 700,000 years ago. And in Africa? In a cave in Swartkrans, more than 270 charred bones are evidence of cooked meat (found with what are likely tuberous roots), proving that the use of fire is in fact very ancient, going back at least 1.5 million years. At the Chesowanja site in Kenya, there are also traces of fire that date back 1.4 million years, but whether this fire was controlled is in doubt; it could also have been a case of using "natural" fire, as opposed to mastering it. The debate is ongoing, but in his 2009 book *Catching Fire: How Cooking Made Us Human*, Richard Wrangham argues that fire was linked to the rapid development of the brain between 1.6 and 1.8 million years ago in *Homo ergaster* in Africa and in *Homo erectus* in Asia, since it gave an advantage to humans who consumed as much animal protein as possible. In addition, mastering fire was also advantageous for artisanal use. This can be seen in at least two *Homo heidelbergensis* sites in Germany: at Lehringen (500,000 years ago) and at Schönigen (400,000 years ago), where the edges of wood spears were burned to increase their hardness. Further, fire was also important for the development of hominin social life, as groups began to gather around the fire and share experiences.

CHAPTER 4

What Obligate Bipedalism Has Made Us

Bipedalism liberated our hands. This biological evolution, shaped also by a cultural evolution, made our hands into veritable tool machines with spectacular abilities, thousands of sensory receptors, and a large part of the brain to guide them. Bipedalism also led to a reconfiguring of the entire body, allowing us to run, and running, in turn, made us lose our fur.

As humans became more bipedal to fully utilize increasingly extensive and diverse terrain and as they completely committed to life on the ground, culture and biology began to influence each other. This coevolution can be seen via the evolution of "prehistoric material culture," the archaeological traces of cultural activities, beginning with stone tools. We begin to see the evolution of the hand along with the brain that guided it toward increased skill and technical diversity.

Before the discovery of tools presumed to be made by *Australopithecus* (at Lomekwi 3), the idea that toolmaking was unique to humans was very old. According to this idea, known as *Homo faber*—Latin for "craftsman"—the fabrication of tools is identified as an absolute singularity, fully distinguishing humans from other hominids and early hominins. The human hand belonging to *Homo habilis*, around 2.6 million years ago, would have knapped stone tools at a site called Gona in Ethiopia, triggering the gradual development of a larger brain necessary to produce and perfect them.

Clearly this idea is a bit of an exaggeration, since the first toolmaker was likely not a member of the *Homo* genus but of *Australopithecus*. This proves once again that human evolution was "bushy." Nevertheless, it's obvious that the fabrication and use of tools exerted considerable pressure on our biology—on the hand in particular—and on our cognition, notably on the parts of the brain that direct the hand.

The Hand, a Veritable Tool Machine

The evolution of lithic knapping technology (the fabrication of stone tools) is reflected in the evolution of the hand. One of the most emblematic results of hominization is that our hand is so distinctively different from that of other hominids. Over time, evolution shrank the human hand—this is especially obvious in the thumb, which is missing a middle bone, but also in the other fingers, which are significantly shorter than a chimpanzee's. Our hand is made up of twenty-nine bones, just as many joints, thirty-five muscles, a vast network of nerves and arteries, and more than a hundred tendons. Furthermore, the bones of our fingers are not curved like those of apes, but straight. Our thumb, the strongest of the fingers, is opposable, and it alone uses nine muscles and the three main nerves of the hand. It's because of these numerous muscular motors and tendon transmissions controlled through nerves, very similar to the strings on a marionette, that our fingers move individually and with dexterity.

Our hand is uniquely different from those of other hominids in its versatility: it can assume a variety of hooklike shapes; it can be a fulcrum and a versatile gripping instrument, with both strength and precision; it also serves as a hammer or a drinking cup, a measuring tool, and much more. Our incredible hand transformed us into a sort of intelligent tool machine, which, with the information gathered by its numerous sensory receptors, can respond to stimuli almost instantaneously.

These microscopic receptors make the hand into an organ of information and communication. The presence of so many nerve fibers—more than seventeen thousand—especially on the palm and tips of the fingers, allows us to have a sense of touch modulated by sensitivity; it's with our hands that we make contact with the external material world. Without even being conscious of it, this provides us every day with millions of pieces of subtle information about the shape, nature, and composition of everything around us, along with the emotional states of our loved ones.

The hand also reflects the incredible extent of our cognitive ability. It's estimated that the movement of the hand requires the use of around 25 percent of the areas of the brain devoted to movement in general, in particular the motor cortex (located in the posterior part of the parietal lobe), which is dedicated to voluntary movement, and a portion of the cerebellum, which triggers coordinated movements. So we can determine that the motor and sensory capacities of the hand contributed to the increase in our cognitive ability and the size of our brain.

Bipedalism, Mother of the Hand and Body

The hand simply would not have evolved without bipedalism. Bipedalism is also the source of many other transformations, of which we should take note: the human body only became adapted to obligate upright locomotion after a series of impressive adaptations in the feet, knees, hips, pelvis, and spinal column, as well as the skull and even the inner ear. A multitude of other biomechanical modifications have affected the tendons in the feet, the abdominal muscles, the development of the buttocks, the strength of the Achilles tendon, the construction of the shoulders, the shapes of the male and female pelvis (in a standing position, they bear all the weight of the organs), and so on.

Bipedalism freed our upper limbs from the locomotive process and enabled them to take on many other tasks. In short, it has utterly reshaped the hominid body. This incredible transformation is not finished, as we are still in the process of adapting to our erect posture. There are still many difficulties created by this posture that we still experience today: the difficult process of childbirth, and the challenge of standing on our feet for several hours without suffering any back pain, for example.

Additionally, our bipedalism allows us to run, which has also required a vast anatomical transformation. When we run, the head cannot bob, as that would be harmful. This requires strong neck and shoulder muscles for support, elongating the silhouette of the upper torso, which is quite different from that of the apes. Have you noticed how their heads seem to sit directly on their shoulders? The human body, too, must remain upright and stable, which the extraordinary development of our gluteal muscles (buttocks) has made possible. And our foot has also been entirely reshaped to store elastic energy in the plantar arch for running.

Running Furless

Moreover, the increasingly frequent practice of running over time explains one of our singular characteristics among the primates: the loss of our fur. Even the hairiest human among us is furless. It's particularly strange because fur has so many important advantages: excellent thermal insulation and protection from abrasions, moisture, the sun's rays, parasites, and pathogens; beyond that, its color, often brown, creates camouflage, and its pattern helps distinguish between members of a species. So how do we explain this peculiarity?

Since skeletons in the fossil record don't offer us any information, we must make conclusions based on what is uniquely human about the functioning of a very precious organ: the skin. Every square centimeter of our dermis on our hands and feet contains at least 600 to 700 sweat glands. We have 180 on the forehead, 108 on the arms, 65 on the back, and so on.

These glands, especially the eccrine sweat glands, produce a transparent fluid on the surface of the skin, unlike other primates, who have a foamy sweat that wets the fur. Fur does persist in humans but only in certain areas of the body—the armpits, the pubic area, and the nipples—and is associated with the apocrine sweat glands, which respond to emotional stimulus (psychological and/or sexual), and not to heat; it has also remained on the head, as hair, to protect us from the sun.

Considering all of this, anthropologists have linked the evolutionary selection of increasingly furless early humans to high temperatures in the savannas, where having a substantial amount of fur would have been a disadvantage, since they are much hotter than the forests. Less furry individuals would have had an advantage during long-distance treks in search of natural resources, and they would have been able to flee from predators more quickly. So decreased fur would have been selected for gradually, while the human body developed, from generation to generation, a more efficient body temperature regulation system.

At the same time, the pale skin that had existed under the fur darkened. As Nina Jablonski, a paleobiologist at Pennsylvania State University, has pointed out in her work on human skin pigmentation, dark skin has increased melanin pigmentation, which protected our ancestors in the tropical regions from ultraviolet rays.

With all of this in mind, let us point out a paradoxical product of evolution that occurred later in humans in the subarctic regions: their skin has become pale. In these countries, it's actually advantageous to have translucent skin, because vitamin D—an essential vitamin for bone health, for example—is only synthesized in the skin under the effect of ultraviolet rays. In the north, pigmented skin is insufficiently penetrated by these rays, so evolution has given these humans translucent skin, beginning in European Neanderthal and Asian Denisovian hominids, and finally in Sapiens populations outside of Africa. Neanderthals were the first to have light skin in Europe, long before the arrival of African Sapiens, whose dark skin would begin to depigment in Eurasia during the last ice age (18,000 years ago), when they were forced to live under the constantly cloudy skies—which diminished their exposure to the sun—of the northern hemisphere. This occurred a very long time after strong African ultraviolet rays had led to the selection of pigmented skin.

It's unclear when exactly this loss of fur occurred in prehumans or humans, and when it gave rise to our current appearance. Since the oldest known human fossils date back to 2.8 million years ago, it's possible that this process had already begun in *Australopithecus* in the savannas 3 million years ago. Does the end of this process coincide with the emergence of *Homo*? That question is up for debate, but certainly the loss of our fur coincides with the advent of hunting.

CHAPTER 5

Hunting Arouses
All of the Senses

The development of obligate bipedalism, tools, running, catapulting shoulders, and furless skin all made hunting possible for our ancestors. This unique and complex combination of features might have been present at the Homo habilis *stage but was surely completely in place from the time of* Homo ergaster, *the first tall human form. The cooperation and coordination necessary for hunting in groups was achieved through hand signals and shouting, which are, together, the oldest form of linguistic communication, and which gradually led to articulate language.*

The concurrent evolutions of our diminishing fur and our ability to run signaled the beginning of hunting live prey. To grasp why, we must understand the huge ecological impact of the hunting activities of our oldest human ancestors.

It started with the evolution of African macrofauna throughout the Paleolithic period. By studying the fossils of large African predators, the paleontologists Lars Werdelin of the Museum of Natural History in Stockholm and Margaret

Lewis from Stockton University in New Jersey discovered in 2013 that, between 2 and 1.5 million years ago, some species of hyenas, saber-toothed tigers, and other giant African carnivores disappeared. They attributed this diminution of large predators to the activity of early humans, who had begun to eat meat in large quantities and used stone tools. It is unlikely that hunters would have attacked these extremely powerful animals directly, but their incessant preying on smaller animals and presence in the ecosystem would have given an advantage to much smaller carnivores who hunted in groups like humans.

In 2004, the biologists Dan Lieberman of the University of Utah and Dennis Bramble of Harvard University wrote an article in the journal *Nature* to suggest that endurance running should be considered a feature of the genus *Homo*. As a consequence, the increased hunting activity of *Homo* species throughout time mirrors our continuing evolution toward running. We know that humans can run at various speeds— for instance, quick sprints when fleeing or attacking, or alternatively slowly and steadily when covering long distances. The adaptation of our legs to these two speeds suggests that our ancestors often explored their environment in groups at a steady pace to find resources or ran quickly to avoid an attack or to perpetrate one.

Running like the San Bushmen

Antelope hunting as practiced by a group of hunter-gatherers, the San, in the desert area of South Africa, evokes these ancient strategies to this day. Since large predators don't hunt during the day to avoid the sun, the San take advantage by hunting antelope in daylight. They certainly can't match them in terms of speed, but thanks to their ability to run and to their sweat system, they have much greater endurance. After a long pursuit, the animals, which are better at short-term speed running, are forced to stop and lie down in order to recover. At this point the hunters move in and kill them without difficulty, even the largest ones. Other hunting strategies, such as cooperative and coordinated hunts, are also based on the fact that when an antelope or deer runs quickly, it will become exhausted if forced to run for long enough.

Additionally, though it's difficult to know for certain, it's likely that birds were the target of stone throwing, the first form of hunting. A well-thrown stone could easily break a wing. We should, once again, point out that the unique construction of the human shoulder enables us to throw an object faster than any other animal: up to 100 miles per hour (160 kilometers per hour) for some baseball pitchers.

Predation, however, could have only developed gradually, alongside a profusion of forms of natural resource collecting. Paleoanthropologists believe that at the evolutionary stage during which humans emerged, these included foraging and opportunistic scavenging. What's unclear is whether hunting was already a normal behavior. Studies of near-current and current hunter-gatherer cultures shows that in tropical ecosystems, foraging provides around 70 percent of dietary resources. From this, we can infer that the hunting behaviors must have developed once complex bipedalism, running, sweating, and the catapulting shoulder were all in place, at the end of the process of hominization, with the first large human species: *Homo ergaster*, the first human species with bodily proportions similar to ours.

The Search for the First (Flint) Blade Runner

Does the archeological evidence confirm this? Matthew Bennett, of Bournemouth University in the United Kingdom, analyzed 1.5-million-year-old *Homo ergaster* footprints discovered at the Ileret site in Kenya between 2005 and 2008. He found that these twenty *ergaster* footprints belonging to two adults and one child possessed an arched foot and employed the mechanical movements necessary for running. Morphologically distinct from the *Australopithecus* footprints in Laetoli from 3.8 million years ago mentioned previously, these footprints in Ileret further indicate for the adults a height of around five feet, six inches (1.75 meters). According to Yvette Deloison, "these footprints are proof of an anatomy specialized for walking and running." Which means that 1.5 million years ago, modern walking and running were already in place, along with a body size sufficient for developing significant physical strength.

Homo ergaster's walking and running are also visible in a structure discovered in 1979 at the Koobi Fora site on the eastern shore of Lake Turkana in Kenya. In this particular hunter-gatherer site, a great deal of the space they occupied, more than 1,000 square feet (100 square meters), was used for the butchering and consumption of meat. This means that *Homo ergaster*'s groups were already sufficiently effective hunters, leading to the establishment of butchering. The co-operation that hunting requires can also be seen at the Ileret footprint site, where the prints seem to have been made by numerous adults, probably males (according to the size of the footprints), moving together alongside a lake where animals would have come for water. It certainly appears as though these tracks caused by walking on a muddy riverbank were made by hunters. Here we can identify one of the central social characteristics of Paleolithic humans: we existed as bands of adult hunters on the lookout for prey and as groups of gatherers, mainly women, elders, and children, who gathered plants to make up a considerable portion of the diet. And we know for sure that *Homo ergaster* consumed meat and knew how to procure it, possibly by employing, like the San, a form of hunting that was increasingly effective because it was well coordinated.

From Body to Hand, to Words, to Language

As we've established, bipedalism freed the hand, enabling us to grasp, evaluate, transform, and make artifacts, which contributed significantly to the development of our cognition. But it also had a huge role in the birth of language, which arose from the need to coordinate groups of (pre)humans, as being able to talk to each other allows us to strengthen our social bonds. And because our hand is capable of transmitting emotion through contact, it was actually one of the first organs used in primate communication. For our distant ancestors (and for other modern-day hominids), it was a tool of social grooming, a form of tactile communication that was essential for group bonding. Hand gestures became symbols of emotions, and thus the first symbolic language was born.

Even today, humans wishing to reinforce an emotion communicated with spoken language reflexively use their hands. We move our hands to stress our meanings. This may be a surprising thought, but think for just a moment about everything that you know how to "say" with your hands: come here, go away, I have a headache, I'm sad, move out of the way, stop, pay attention, beware. Whether this type of communication is complex and highly codified, as with Italians (world famous for their complex hand signaling while speaking; a Roman inheritance, probably linked to the need for communication during antiquity [8th century BCE to 6th century CE], when different languages were spoken to the slaves brought to Italy), or simpler, it's clear that it is both spontaneous and organized, like spoken language.

In 2014, having analyzed all of the available research on the subject, Catherine Hobaiter and Richard Byrne of the Primate Research Group at the University of St. Andrews, in Scotland, determined that the tendency to use the hands, body, and voice simultaneously to communicate an observation (for example, to say "Watch out for the snake!") is present among all great apes. This, along with the fact that these actions are often spontaneous and unconscious, proves that this form of symbolic communication is very old. Among primates, symbolic language began with the body (including facial expressions to threaten or to express submission, for example) and extended over time to include the hand and voice (to signal from afar or to alert).

At some point, progressive complexification resulting from the introduction of increasingly articulate sounds led ultimately to a separation between manual symbolic language and acoustic symbolic language (voicing of words). As a matter of fact, acoustic symbols (words) can be added nearly at will, while it is much more difficult to add manual symbols, as we have only two hands and ten fingers, making it much more difficult to multiply hand gestures and establish their meaning. Once separated from the language of the hand, resonant words (vocabulary) increased and their sequences organized to add supplementary meaning (grammar). It's undeniable that the combined body-hand-voice language typical among primates has become somewhat disassociated among humans; we have body language, manual language, and spoken language, and we are the only primates who know how to use them independently of each other.

Talking with the Hand

According to our hypothesis, manual communication led to oral communication, at which point the hands, body, and language began coevolving. When was this self-reinforcing loop triggered? According to the model by Leslie Aiello and Robin Dunbar, a *Homo ergaster* devoid of language would have devoted more than 25 percent of its time to social grooming. The complex cooperation and coordination involved in a group's harvesting, foraging, hunting, and butchering activities makes it more likely that clans of *Homo ergaster* already had the capability to communicate verbally.

The first language undoubtedly took the form of code rather than true spoken language, considering it arose from the animal communication used by prehuman hominids. On the other hand, our contemporary language faculties are elaborate but also abstract, as they allow us to communicate not only about concrete objects, facts ("This is water"), and situations ("Be careful, there is a snake"), but also about imaginary objects, ideas, and events (mathematics or mythology, for example). Over thousands of years, the ability to speak has diversified into a great number of different languages (today there are more than seven thousand), numerous coded communication techniques (such as computer languages and telephone numbers), abstract forms (such as mathematics), and even many phonatory forms (shouts, whistling, songs). A high level of diversification is an indication of great antiquity.

But When Was Articulate Language Born?

So can genetics come to our aid in order to evaluate the age of articulate language? Since the late 1990s, we've known of at least one gene, shared with many vertebrates, that plays a role in communication, for example in the development of bird song. In humans, this gene, FOXP2, plays an important role in articulated speech. This is why it is sometimes misidentified as "the language gene." Located on chromosome 7, this gene was initially identified as the cause of a speech disorder in a family that presented with a mutation. We know that the version of this gene present in chimpanzees differs from ours. We also know that Neanderthals (350,000 to 40,000 years ago) carried the same version we do. From this, we can conclude that the common ancestor of Sapiens and Neander-thals possessed our version of the FOXP2 gene, which has existed in this form for at least 600,000 years, back to the time of *Homo heidelbergensis*, our common ancestor. It would be interesting to know if this form of the gene FOXP2 was already present in early *Homo*, notably in *Homo ergaster*, but it's impossible to go any further back in the genetic record.

A much more realistic approach would be to confirm in *Homo ergaster* the presence of the complex phonatory system necessary for articulated language. This was done in the case of *Homo neanderthalensis* through the discovery in the Kebara cave in Israel of a hyoid bone some 60,000 years old. This horseshoe-shaped bone is indeed the centerpiece of the entire oral-laryngeal-pharyngeal edifice. However, it is neither articulated nor attached to any other bone, so in modern humans it can only be studied by X-ray or dissection. It is extremely rare in archaeology, and since it's nearly impossible that a hyoid bone fossil could have survived after millions of years, we cannot know for sure what a phonatory system would have looked like in our earlier ancestors, such as *Homo ergaster*.

We must therefore be satisfied with attesting to the presence of a phonatory system in these ancestors based on an observation of the position of their larynx, since this is the result of the evolution of the entire base of the skull. Indeed, for *Homo habilis*, who already had an elongated neck, the position of the larynx is comparable to ours. Although the oral cavity is smaller, it is deep and typically associated with greater flexibility of the tongue and lips, features that now allow us to sing as well as to speak. Thus, it is most likely that *Homo habilis* already had an advanced phonatory system, and earlier hominins did too.

Therefore, because we can identify that a mode of communication involving the body, hands, and vocalizations existed in our common ancestor with the great apes and confirm the existence of a material culture in some *Australopithecus* more than 3.3 million years ago, it seems plausible that the first form of articulated language emerged before the Lomekwian stone tools, that is, between *Ardipithecus* and *Australopithecus*, between 4 and 3.5 million years ago. This means that during this time, prehuman forms did indeed "proto-speak" (practiced a kind of proto-language), though this seems very difficult to prove.

The casts that paleontologists make of the inside of the skull, however, might be able to help a little. Fossils from across the *Homo* genus attest to the presence of a cerebral asymmetry that, in *Homo sapiens*, has become especially pronounced. The left part of our brain is a not a reflection of its right side, and this is an ancient phenomenon, as examination of the cortical surface of *Homo habilis* offers evidence of the existence of the Broca's area as far back as 2 million years ago.

Identified as the motor area responsible for speech production by the French physician and anthropologist Paul Broca (1824–80), this region is located on the left frontal lobe of the cerebral cortex. Since Broca's area is an important part of the cortex, its evolution, like any complex biological system involving a very large number of genes, is estimated to have taken hundreds of thousands of years, rather than tens.

In this case, the hypothesis that the evolution of a proto-language in *Australopithecus*, before the appearance of the first tools at Lomekwi—at least a half million years older than the pre–*Homo habilis* jawbone LD-350-1—is, at the very least, plausible.

We don't know when the first proto-language appeared, but one thing is certain: around 2 million years ago, *Homo*'s biology provided it with all of the means—complex bipedalism, tools, language—necessary for an increasingly social exploitation of the land. And it began to use these means to conquer the entire planet.

CHAPTER 6

The First Conquest
of the Planet

Before Sapiens emerged, numerous migrations out of Africa occurred, one after another, though we don't know exactly when or how. Rare direct fossil evidence of this, along with numerous indirect ones in the form of stone tools, indicates that the expansion of Homo *began well before 2 million years ago. The result of many migratory movements and crossbreeding, these first Eurasians then mixed with Sapiens when they, too, expanded out of Africa.*

The migrations by African *Homo* are among the largest geological and historical events in the history of the planet Earth (Figure 10). Humans stand apart from all other animals in that their ancestors unshackled themselves from the tropical forests to adapt to the African savannahs, the semidesert landscapes, the Mediterranean, and eventually to all climates, including the Arctic.

These numerous adaptations of *Homo* occurred in large part after they moved out of Africa. From archeological and paleontological data, prehistorians have identified four main migrations. The first occurred more than 2 million years

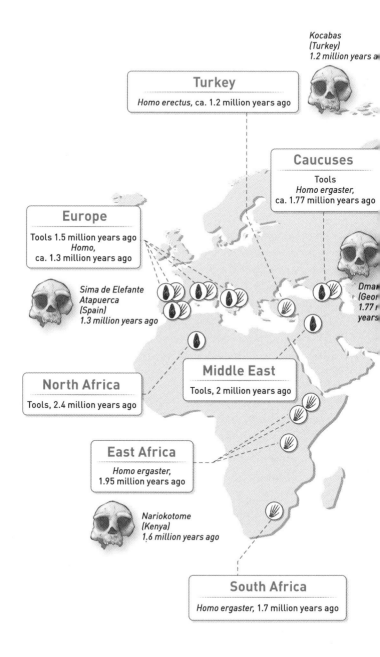

Kocabas
(Turkey)
1.2 million years a[go]

Turkey
Homo erectus, ca. 1.2 million years ago

Caucuses
Tools
Homo ergaster,
ca. 1.77 million years ago

Europe
Tools 1.5 million years ago
Homo,
ca. 1.3 million years ago

*Sima de Elefante
Atapuerca
(Spain)
1.3 million years ago*

*Dma[nisi]
(Geor[gia)]
1.77 [million]
years [ago]*

North Africa
Tools, 2.4 million years ago

Middle East
Tools, 2 million years ago

East Africa
Homo ergaster,
1.95 million years ago

*Nariokotome
(Kenya)
1.6 million years ago*

South Africa
Homo ergaster, 1.7 million years ago

Figure 10

The first conquest of Eurasia

The first migrations out of Africa can be seen from indirect evidence, such as stone tools and other artifacts, and from direct evidence, fossilized human remains. The oldest migrations took place toward warm climates—it was later that these humans reached temperate ones.

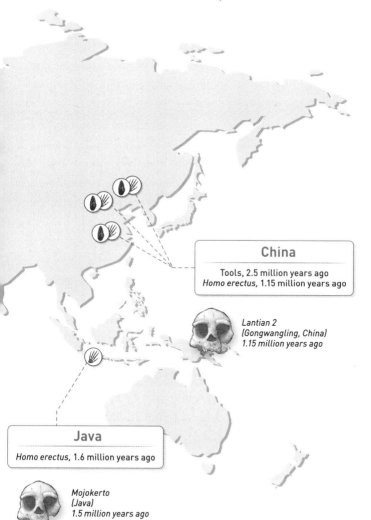

China

Tools, 2.5 million years ago
Homo erectus, 1.15 million years ago

Lantian 2
(Gongwangling, China)
1.15 million years ago

Java

Homo erectus, 1.6 million years ago

Mojokerto
(Java)
1.5 million years ago

ago, the second some 1,400,000 years ago, the third around 800,000 years ago, and the fourth around 200,000 years ago. Through the analysis of genetic data, they have also identified one final migration, which occurred around 70,000 years ago. We believe, however, that starting with the first migration, Africa, like a cauldron under pressure, continually emitted small groups of people over time. It is also possible (and proven for the last few millennia) that movement of populations have occurred also in the other direction.

The standard is to group all of the human figures that left Africa, some 2 million or more years ago, under the term *Homo erectus*. This is the name historically used to refer to the Asian form of *Homo ergaster*. Evidence of hominids outside of Africa has been found in Asia (Georgia, India, and China), then in southern Europe (Italy and Spain).

There are two kinds of evidence of these migrations out of Africa: indirect, meaning traces of human occupation seen only in the presence of stone tools and artifacts (Figure 11), and direct, in the form of fossilized human remains. In the last decades, three examples of indirect evidence (knapped stone tools) have been identified in subtropical China: the oldest, aged 2.5 million years, at Longupo, in the Chongking Province, another one dated to 2.2 million years ago at Renzidong in the province of Anhui, and finally one 2.12 million years old at Shangchen in central China. Then, after a long time without hints of movement, from about 1.7 million years ago, numerous stone tools start to appear in temperate China. The presence of these old traces of human evidence calls attention to another site found in the 1990s at Yiron,

Years ago

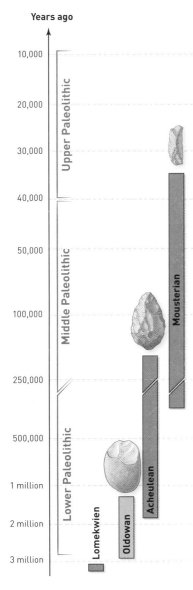

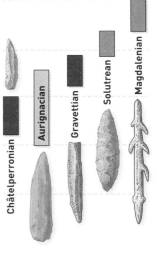

Figure 11

The main material cultures
The evolution of prehistoric cultures is perceptible above all in stone tools. The first well-established material cultures, Lomekwian and Oldowan, can be seen in Africa. The Acheulean culture spread out of Africa, into Europe and parts of Asia. The Aurignacian and the Gravettian cultures developed before the last ice age that covered Europe. The wonderfully technical Solutrean culture corresponds to this last glacial maximum, and the Magdalenian follows it.

Israel, dated to 2.4 million years ago, which was discovered by Avraham Ronen of Haifa University. This incredible discovery proves once again that the Levant corridor (the land running today from the Sinai Peninsula to Lebanon) has always been the main pathway toward Asia.

Which *Homo* might have left these trace fossils in China? For years (before the discovery of indirect evidence of stone tools older than 2 million years) we thought that it was *Homo erectus*, the Asian form of the African *Homo ergaster*, but this *Homo* is known to have existed in Africa later than 2 million years ago. Therefore, some paleoanthropologists propose that groups of *Homo habilis* indeed were the first to follow herds of animals out of Africa. Possibly, the next decades of fieldwork might bring big surprises.

The oldest direct evidence of *Homo* out of Africa was found in Dmanisi, Georgia. These fossils, dating back to 1.8 million years ago, included several complete skulls, mandibles, and other postcranial (body) remains. Their great diversity has stunned researchers to the point that in order to differentiate them from the African fossils of *Homo ergaster* already known to be from the same time, some paleoanthropologists proposed a new species—*Homo georgicus*—causing much debate in the scientific community. For some prehistorians, the lesser cranial capacity of these specimens, compared to *Homo ergaster*, suggests that they actually belonged to more primitive creatures, closer to *Homo habilis*. This assumption is reinforced by the fact that the artifacts found at the site of Dmanisi did not contain any bifacial stone tools (symmetric tools also called hand axes: see Figure 11). In fact, the hand axes are a key cultural trait of *Homo ergaster*, because they appeared in Africa only 1.8 million years ago. Hence the morphology of some of the fossils of Dmanisi give support to the paleoanthropologists who believe that *Homo habilis* was the first *Homo* out of Africa and those who believe they could also be a geographic kind of *Homo ergaster*.

What we do know for sure is that some form of *Homo* entered Asia more than 2 million years ago. This is proven in both south China and on the islands of southeast Asia. Several specimens were found in the hot climates of these regions, which suggests that the tropical humans who migrated from Africa more than 2 million years ago first settled in warmer regions before reaching temperate zones. In Indonesia, for example, the first traces of tool fabrication and human remains date back to around 1.6 million years ago, in Mojokerto and Sangiran on the island of Java; in the same area, at Trinil on the same island, fossils from about 800,000 years ago have been discovered.

Homo erectus Visits Europe Without a Visa

The earliest indirect hints of human presence on European soil date to around 1.5 million years ago. They're indirect traces more than anything else—lithic tools (non-bifacial) or marks left by tools on bones. The oldest sites are predominantly in the southern part of the continent, at Pirro Nord, for example, in southern Italy, but also in Spain, in Andalusia, at Fuenta Nueva and Barranco León, and near Atapuerca at Sima del Elefante. Around 1.4 million years ago, the first direct evidence appears: a human tooth at Barranco

León and a mandible dated to around 1.2 million years ago in Atapuerca.

Starting around 1 million years ago, humans began multiplying in Europe. It seems that the part of Europe located north of the fortieth parallel (the Madrid-Sardinia axis) wasn't occupied until almost 800,000 years ago, a period in which a dozen stunning footprints were preserved in mud, near an ancient estuary located near Happisburgh, in eastern England. These footprints, the oldest of any found in Europe, tell us about the size of these first northern Europeans. The length of their strides, the depth of their prints, and the size of their feet indicate at least five individuals measuring from three feet (90 centimeters) to five feet six inches (1.7 meters) in height. These five, comprising adults and children, crossed the land that at that time connected England to the rest of Europe.

Around 700,000 years ago, the sudden appearance of evolved hand axes (more symmetrical and slimmer than their earlier counterparts) in Europe seems to reflect the arrival of a new human that has been identified as *Homo heidelbergensis*. Note that bifacial tools—hand axes—have been found outside of Africa, notably in the Middle East, at the Oubedija site in Israel, dating to around 1.4 million years ago. Associated with a material culture called Acheulean (named after the Saint Acheul site in France, where the first hand axes were discovered in 1859). While some *Homo* left Africa, *Homo ergaster* evolved there to become *Homo heidelbergensis*, the presumed ancestor of Sapiens and European Neanderthals. Once in Europe, *Homo heidelbergensis* probably reproduced with descendants of the first local European humans (well known to us from Spanish sites). Then, through the phenomenon of genetic drift, in which, due to partial isolation, certain genes are selected over time, its evolution led, after several hundred thousand years, to the emergence of *Homo neanderthalensis*. It is also likely that *Homo heidelbergensis* made headway into Asia, where it would have given rise to a third form of humans, the Denisovans, whose existence was revealed first to us only through genetics, since the only fossils we've found are a tooth; a phalange bone from the hand, containing DNA; and small bone fragments—all from Denisova cave. Thanks to a 2019 study, the Denisovans are also known today from a partial jawbone found in Tibet.

The Sapiens genome revealed that our ancestors were crossbreeding. We know that modern-day Eurasians' ancestors crossbred with Neanderthals and Denisovans, who had also interbred. Today, the Sapiens bearing the strongest traces of this crossbreeding are the First Nations people of Australia and the Melanesians, whose genes still contain 4 to 6 percent Denisovan genes, compared to about 1 percent for Sapiens living on the Asian mainland.

CHAPTER 7

And *Homo sapiens* Emerged . . .

It was thought that Sapiens evolved from Homo heidelbergensis *in East Africa, but it has become clear that this process happened throughout the entire continent. Sapiens then began a process of population growth, which pushed them out of Africa. Early* Homo sapiens *lived in hordes (band society), which profoundly impacted their psyche, giving them a strong sense of group belonging, a great talent for empathy, and a certain tendency toward altruism.*

So far we have focused on the *Homo* genus, but everything we have discussed is also about Sapiens, since it concerns the human lineage. Now let's talk about our subject: *Homo sapiens*. Like Neanderthals in Europe, *Homo sapiens* presumably arose from *Homo heidelbergensis*, but in Africa. It's there that Sapiens first appears in the fossil record, even though its origin is unclear. We think that it is the case, but don't know for sure if Sapiens actually derived from *Homo heidelbergensis*, which descended from *Homo ergaster* and *Homo erectus*, the bearers of Acheulean material culture, which is found all over Africa. However, the lack of fossils—especially for a gap from 600,000 to 260,000 years ago—muddies the possibility of verifiably linking *Homo heidelbergensis* to ancient human species and *Homo sapiens.*

Until recently, the fossil record for Sapiens has been meager: apart from the two 196,000-year-old skulls Omo-1 and Omo-2, found in the Omo Valley in Ethiopia, we've had little beyond the incomplete 285,000-year-old skull found in Florisbad, South Africa, and some skull fragments and jawbone from 125,000 years ago, discovered in caves near the Klasies River, also in South Africa. They've proved the existence, in Africa, of a human form with a high forehead like ours. Recently, though, new discoveries have made it much more complicated.

In 2018, one half of a left-side Sapiens maxilla bone from between 177,000 and 194,000 years ago was discovered at Mount Carmel, Israel, and called Misliya-1. This proved that early *Homo sapiens* left Africa at least 100,000 years earlier than previously thought. This was a surprise: before the discovery of Misliya-1, the dominant idea was that only a few *Homo sapiens* had dared to step outside of Africa around 100,000 years ago and that most had waited before venturing out until after 70,000 years ago.

In 2004, there was a reexamination of the site at Jebel Irhoud, in Morocco, North Africa. The fossils from this site have a long history. They had been first unearthed in the 1960s, along with stone tools made using the Levallois technique, which at the time had been exclusively associated with Neanderthals. This flint knapping technique, which allowed for more edge lengths per volume unit than previous techniques, is a hint of an advanced cognition. Thus, the bones had also been first attributed to Neanderthals, but they were later qualified as "Neanderthaloid," because certain fine anatomical details of their face and the back of their skulls distinguished them from European Neanderthals. Finally, after the discovery at the same site of two partial mandibles in the late 1980s, paleoanthropologists distinguished them from Neanderthals and started to call them "archaic Sapiens" or "ancient Sapiens," since they were thought to be 150,000 years old.

This confusing situation meant that the status of Jebel Irhoud's fossils had to be clarified. Starting in 2004, a team led by Jean-Jacques Hublin of the Max Planck Institute for Evolutionary Anthropology in Leipzig, Germany, undertook another study of the site and the fossil record. In 2017, the features of the fossils were analyzed using modern techniques. Evidence emerged showing that the humans of Jebel Irhoud were actually closer to *Homo sapiens* than to any other species. Simultaneously, the redating process, done using a series of independent dating techniques (on the tools accompanying the fossils and the soil strata that held them) produced a staggering result: they were 315,000 years old.

This new date was shocking because *Homo sapiens* was thought to have emerged in eastern and southern Africa exclusively, so the Jebel Irhoud discovery led, at first, to a great deal of skepticism among a number of prehistorians. But one fact remains: it has changed the view we've held concerning the emergence of *Homo sapiens*. The existence of these ancient Sapiens also in North Africa renders the notion of a cradle of Sapiens in east and south Africa obsolete, and suggests a Pan-African presence instead.

This view was supported in 2018 by a large international consortium of leading scientists. After closely examining all of the climatological and cultural data as well as African variability concerning the period during which Sapiens emerged, this consortium concluded that human evolution in Africa was multiregional. Since broad evidence of the emergence of current human traits is observable, it is clear that pockets of populations bearing a mix of features (some of them more like Sapiens and others resembling those observed in archaic humans) existed in different regions of Africa at various times. As Sapiens emerged, "the evolution of human populations in Africa was multi-regional; our ancestry was multi-ethnic; and the evolution of our material culture was, well, multi-cultural," summarized Eleanor Scerri (at the time, a visiting scholar at Oxford University; now at the Max Planck Institute for the Science of Human History, in Jena, Germany), one of the authors of this idea.

Sapiens, A Longtime Networker

This conclusion is of major significance, since it debunks the original hypothesis about our origins, with East Africa being the origin of humanity. Africa is fragmented: large rivers, equatorial forest, and expansive deserts imply that circulating from one region to the other was not always easy. Nevertheless, a network of connected prehuman and human habitats has clearly always covered the African continent, often under the influence of climate changes. This makes the increasing human migration out of Africa more than plausible, since the northern and eastern parts of this network of ecosystems likely extended at times through the Levant and southern Arabia. Thus, like *Homo ergaster* and *Homo heidelbergensis*, *Homo sapiens* unintentionally left Africa more than 200,000 years ago. Remember, before Misliya-1, Sapiens was thought to have left Africa a mere 60,000 years ago.

So how should we interpret these discoveries? As we've already stated, it is most likely that *Homo sapiens* is descended from *Homo heidelbergensis*, but it's clear that the evolution of Sapiens in Africa remains very enigmatic. More than anything else, what stands out is the emergence throughout the continent, between 300,000 and 400,000 years ago, of a new type of material culture, characterized by a tendency toward miniaturization and a generalization of the Levallois technique of knapping sharp blades that we've already mentioned.

The evolution from *Homo heidelbergensis*, a human form with a large brain (not just in proportion to the size of its body), indicates the complex changes that will take place later in Sapiens, particularly those involving behavior and the exchange of material techniques (and probably genes) across Africa. Due to the scarcity of fossils from between 1 million years ago and 600,000 years ago, we unfortunately do not know much about how this evolution actually occurred.

A Cognitive Revolution?

Homo sapiens's behavior is singular: we have flooded the entire planet and profoundly transformed most of its ecosystems, even influencing the global climate. But where do the origins of this behavior lie? From our point of view, nothing biological could possibly explain the singularity of Sapiens.

One theory, expressed in the book *Sapiens* by Yuval Noah Harari, is that there must have been a "cognitive revolution" that distinguished Sapiens from other humans. We disagree with this assertion because Neanderthals and Sapiens, living at the same time, had similar technological skills, among them the advanced Levallois technique and material cultures. Both species spoke and used symbolic written languages (in the form of ornaments and paintings). Even though a number of aspects of Neanderthal's body—a stockier appearance, flatter face, and an elongated skull—were different from Sapiens's, the cerebral volumes of the two species were comparable, with a slight advantage for Neanderthal. It was only much later, when Sapiens had already conquered the planet, that the development of social culture led to a remodeling of its brain.

A comparative study of the genomes of modern-day Sapiens and Neanderthal has shown that Sapiens possess about a hundred genetic mutations. The effects can be seen throughout the body—especially in the skin, immune system, and musculature. So can we infer a biological superiority for Sapiens over Neanderthal? No—this would be a prejudiced conclusion. Since the two species lived in hunter-gatherer groups and had similar environments (the Middle East and Europe), they had, according to archeological evidence, comparable success with foraging and hunting, and frequently interacted, trading materials and exchanging genes. As a matter of fact, the modern European genome contains 1 to 3 percent of Neanderthal genes, which, considering the genetic erosion in the last 2,500 generations since the disappearance of Neanderthals, is quite a lot.

Grow and Multiply

Therefore, what truly separates Sapiens from Neanderthals is our behavior in relation to nature: our ecological behavior. Neanderthals behaved just like any another predator, removing from the environment only the necessary resources for them to survive, which always amounted to much less than Nature could provide. But Sapiens has a totally different behavior. As *Homo sapiens* hordes multiplied over time, they continued to take more and more resources from the environment, endangering one species after another, a process that—unfortunately for life on Earth—has never ceased.

Occupying more and more space in nature, they caused the extinction, in Eurasia and then in the Americas, of most large species of mammalian herbivores (mammoths, bison, cave bears, woolly rhinoceros), predators (saber-toothed tigers, giant hyenas, cave lions) and other hominids (Neanderthals, Denisovans, and "Flores Man"—a small species of human discovered on the island of Flores in Indonesia). These extinctions occurred primarily because the habitats of these large species dwindled or disappeared (Figure 12). The simple presence of *Homo sapiens* in a habitat always changes the living conditions, because they always multiply their numbers.

This capacity for growth has a cultural and social origin: Sapiens invests more time in raising its offspring than any other species. We have a reliable indication of the ancient nature of this behavior in the notable difference between the growth rates of Neanderthal and Sapiens babies: while the former was practically an adult at twelve years old, the latter continues to grow and learn for much longer. Sapiens's

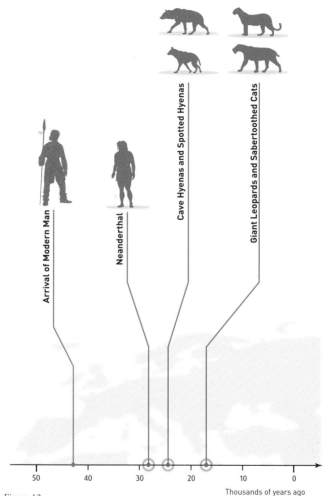

Figure 12

The extinction of large predators in Europe since the arrival of *Homo sapiens*

While Neanderthals have always remained in ecological balance with nature, *Homo sapiens*, in constant demographic growth, has always taken more from the environment. As a result, over thousands of years and throughout the planet, Sapiens has triggered the disappearance of all other large carnivores, including all other human forms (Neanderthals, Denisovans, "Flores Man").

period of growth has lengthened as life expectancy and cultural complexity have increased as well. We know now that the brain doesn't stop developing at age twenty-five—the Sapiens learning period can last up to thirty years, or even well beyond that, since there is great brain plasticity throughout a human's entire life.

Thus, in our opinion, it isn't biological developments that created the singular evolutionary history of Sapiens but its social and cultural complexity, which led to biological changes (increased size of the cerebellum and increased connectivity between neurons). This is confirmed by research in the field of neuroscience, which shows that while genetic heritage plays a fundamental role in the development of young Sapiens, the phenomena modulating the expression of these genes—epigenetics—are crucial. As a matter of fact, the development of a young human and its epigenetics are modulated by its living conditions—essentially, the nourishment it receives and the environment (society) that surrounds it. This means that, for Sapiens, society quickly became much more extensive. Today, each one of us may have tens of friends and relatives (much more than a horde contained) and hundreds of virtual friends on social networks.

The Horde, Origin of All Societies

The first form of society in which Sapiens lived is the same as that of many other species: the horde, or band society, a wandering band of cooperating hunter-gatherers isolated in the wilderness. Cooperation is necessary for hunting and foraging, for the group to prosper and, over time, for the transmission of useful knowledge across generations. Just as there are biological traits, there are also communicable cultural traits that circulate within the horde, through imitation and language, which form its tradition. And as surprising as it might seem, cultural traits that arose during the long evolutionary stage of horde society are still with us today.

The most striking of these, and the most vital, is the sense of belonging to a group. The emotional desire to be a part of a group has long been considered by psychologists to be a basic human need. This trait has been imprinted on us by a selective pressure: while a horde can survive isolated in the wild, an individual will almost certainly die. So, as humans have been selected, over millions of years, by their genes and cultures, to survive, they've also been made to identify with the group on a deep level for a greater chance at survival. Buried in our subconscious, this desire to belong endures within us, whether it's to be part of a tribe, a company, a community, a country, or even to humanity as a whole.

This behavioral trait is the indirect product of human empathy—the capacity to understand the emotions of another. Thanks to our elevated cognition, Sapiens is indeed able, beginning at a very young age, to become self-aware: some babies are able to recognize themselves in a mirror at eighteen months. Armed with this self-awareness, every human has a capacity to identify himself and then use that identity to figure out how to behave appropriately within a group.

Our talent for empathy and our strong determination to belong partly explain the almost universal tendency of humans to protect the weak, thereby protecting the group as a whole. It's also reflected in the tendency to protect "women and children," the future of the group, in dangerous situations; this is a form of horde altruism for which we can easily find an evolutionary interpretation. While a strong man and able hunter can easily be killed by a bull's horn, for example, the loss of a woman is actually much more serious, because it noticeably diminishes the reproductive power of the group.

The altruistic urge to protect also extends to the elderly, who pass on their experiences. We have a great deal of archaeological evidence of this, from the Neanderthals of Chapelle-aux-Saints in France and Shanidar in Iraq, who suffered from debilitating diseases but nonetheless lived to advanced ages. We also have evidence of Neolithic Sapiens that were trepanned using stone tools, to relieve them of strong headaches, beginning roughly 7,000 years ago. The human inclination toward mutual aid, to which we owe the practice of modern medicine, is likely very old. One piece of evidence supports this notion in particular: Miguelón. This half-million-year-old Spanish fossil of *Homo heidelbergensis* was discovered at the Sima de los Huesos site in the province of Burgos, in Spain. While the fossil reveals a very serious bone disease, we can tell that this individual survived into adulthood, which would not have been possible without care from the clan.

Empathy, an Ancient Human Ability

The human capacity for empathy is much older. We mentioned that our evolution toward a body capable of running has deprived humans of fur. Our human evolution also led to the development of the sympathetic nervous system. This is the autonomous part of our nervous system that controls organic automatisms and induces a series of involuntary physiological reactions (heartbeat, emotional sweating) that alter facial expressions when an emotion occurs. The mouth, for example, may open involuntarily in astonishment, the eyebrows might frown, the forehead might wrinkle in anger, the facial features might stretch from anxiety, or the capillaries in the cheeks might expand as a result of anger, social fear (shyness), or sexual attraction. This expansion of the capillaries—and the redness it causes—is particularly visible in humans whose skin is quite translucent, an almost public manifestation of emotions. In *The Expression of the Emotions in Man and Animals*, Charles Darwin identified in these physiological reactions involuntarily emotional communication, thus reinforcing our empathy as one of the most human traits.

Psychologists consider this trait to be part of the fight-or-flight response of any animal in a stressful situation: the adoption of either aggressive and combative behavior or, alternatively, avoidance or escape. In *Homo sapiens*, with no fur, these responses occur most in social life: the fight may, for example, manifest itself in aggressive and argumentative behavior; the escape in social withdrawal or, more subtly, in friendly behavior in order to bring into play the social reflexes of a seemingly dangerous person. By blushing, a person unwittingly signals a great emotion: he or she may either "blush with anger" or, on the contrary, be shy, which will inevitably create the disarming impression of fragility and sincerity, which triggers both the reflex to protect the weak and the reflex to trust. But the role in social life of physiological emotional reactions could be as old as the loss of our fur, thus going back to *Homo ergaster* or even *Homo habilis* several million years ago. The beginnings of human empathy could also be this old, since empathy is present in groups of other primates, especially in chimpanzees.

Our capacity for empathy is part of our ability to trust—and mistrust. Humans devote a significant part of their attention to identifying cheaters. What is a cheater? Among chimpanzees, as the primatologist Frans de Waal of Emory University has shown, a cheater is an individual who does not return favors in proportion to those he has received. According to his observations, a chimpanzee accumulating profits without giving back will experience more and more aggression. If we accept chimpanzees as a similar model of primitive hominin behavior, then at the other end of the evolutionary trajectory of our lineage, we must recognize in our daily behavior that we have the same tendencies. As long as we feel safe, we can easily attack anyone perceived to be cheating, whether by not following a rule or by not sharing.

Where Does Sapiens's Talent for Multiplying Come From?

All of the above traits contribute to the collective psychology of a horde, with Sapiens or otherwise. We think that Sapiens hordes must have possessed something special, beyond these traits, that might explain its spectacular ability to multiply across the planet. But archeological evidence, which is similar to that for Neanderthals of the same period, doesn't offer us anything. We believe the distinctive behavioral trait at the root of Sapiens's success may be linked to division of tasks between the members, particularly between the sexes. As a matter of fact, all ethnographic observations indicate a tendency within hunter-gatherer Sapiens societies to separate tasks according to gender. This separation was reinforced in the Neolithic period (roughly 12,000 to 1,000 years ago) and remains with us today in most societies.

CHAPTER 8

The Spread of *Homo sapiens* Over the Entire Planet

More than 135,000 years ago, the first Sapiens left East Africa to venture into the Arabian Peninsula and then into the southern parts of Eurasia. They arrived in Australia 65,000 years ago and in China more than 100,000 years ago, and for a long time, multiplied their numbers in these hot climates. Then, starting about 60,000 years ago, after mixing with non-Sapiens populations already present in Eurasia, they began moving farther north, entering Europe only around 43,000 years ago, and America 20,000 years later.

*H*omo sapiens's migration from Africa distinguishes our species from all of the other hominins, because at some point, Sapiens stopped being subservient to a particular ecosystem (a tropical one) and have now conquered all of the biomes on Earth, including Antarctica, even modifying Earth's climate. We're also on our way to conquering space. Even though it seems to have happened slowly according to our notion of time, Sapiens's expansion out of Africa was an explosive event on a geological time scale.

We don't know when this major event was triggered, but the half of a Misliya-1 maxilla bone tells us that *Homo sapiens* was already living outside of Africa, in the Levant, more than 200,000 years ago. Later, their descendants, along with the newly arrived Sapiens, found themselves in contact with Neanderthals, who had expanded to the Middle East from Europe. Numerous pieces of material culture left behind by these two human species are identical. The genetic analysis suggests that genetic exchange had begun to take place at least 100,000 years ago.

At around this time, groups of Sapiens left behind numerous well-preserved fossil sites in the Levant, most notably in the caves of Skhul in the Carmel valley and Qafzeh, near the city of Nazareth. According to the traditional narrative, these ancient Sapiens from the Levant would have been blocked in their progress north toward Europe by the presence of Neanderthals. We believe, however, that their interactions with these new Neanderthals who came from the Levant were likely peaceful, at least in part, since the genetics indicate that at around 100,000 years ago, Neanderthal and Sapiens interbred. It's clear that, more than anything else, it was the cold climate, not the presence of Neanderthals, that blocked Sapiens's progress north for so long.

Since there were probably members of our species in the Levant at least 200,000 years ago, even 300,000 years ago, as evidenced by the studies of the human fossils found at Qesem (north of Israel) and Zuttiyeh (near the Carmel valley), it is clear that the expansion of Sapiens out of Africa took place over a long period of time. A team led by Katerina Harvati of the University of Tubingen in 2019 suggested an expansion of these ancient Sapiens in southern Europe, particularly in the Balkan peninsula at the Greek site of Apidima, around 210,000 years ago. But today the most robust and extensive paleoanthropological and chronological data show that these migrations began toward the south, since tropical Sapiens could go there easily. One nebulous but important piece of evidence suggests a "launch date for the conquest of the planet": traces of human settlement in a shelter in Jebel Faya, located near modern-day Dubai, in the Arabian peninsula. The strata in the area of this collapsed shelter date back to around 125,000 years ago and contain bifacial cutters, scrapers, and other shards, which were fabricated using knapping techniques typical of Sapiens, who then populated East Africa.

They Went Through Arabia

This brings us to the hypothesis that African Sapiens reached and occupied the Arabian peninsula more than 125,000 years ago. It should be noted that the northern expansion of the monsoon zone transformed the peninsula, now known for its aridity, into a verdant expanse through which herds of large herbivores passed, between 160,000 and 150,000 years ago, and again between 130,000 and 75,000 years ago. However, 125,000 years ago, Earth experienced a warm period, at the beginning of which global sea levels rose very quickly, to a point around thirty feet (ten meters) higher than current levels. Before this oceanic rise, the journey from Africa to Arabia, and from Arabia to Asia, could only be made by crossing the Strait of Bab-el-Mandeb separating Djibouti and Yemen, which today is only 20 miles (30 kilometers) wide and no deeper than 100 feet (30 meters), and then the Persian Gulf, which is, on average, only 160 feet (50 meters) deep.

It's this last depth that tells us when the first Sapiens were able to pass en masse from East Africa to Asia, through the southern part of the Arabian peninsula: prior to 135,000 years ago. The climate at that time was cold, so the sea levels were much lower and the climate much more humid in the Sahara and in the Arabian peninsula. Conditions like these certainly would have favored the expansion of Sapiens beyond East Africa into Asia, because a band of steppes connected a green Sahara to the Indian subcontinent, through Bab-el-Mandeb and the Persian Gulf. Sapiens from the Levant were also able to move to the south of modern-day Iraq and Iran to join those who had passed through the Arabian peninsula.

From Africa to Australia on Foot

We believe that the Sapiens who reached Asia in the greatest numbers, some 135,000 years and earlier, came from East Africa. Accustomed to the tropics and therefore unsuited to the cold, they could only have progressed into the southern parts of Eurasia, and of southeast Asia until they reached Australia, since the route was tropical and subtropical. When did this first wave arrive in Australia? Three Sapiens skeletons from around 40,000 years ago, discovered on the shores of Lake Mungo, in New South Wales, Australia, provided the answer.

But in 2017, there was a discovery that unsettled this chronology. While excavating a stone shelter in Majedbebe, in the far north of Australia, a team led by Chris Clarkson, archaeologist at the University of Queensland, found stone tools, most notably a number of choppers, which had been stored and buried against a rock wall at the base of the shelter. The researchers then dated them using optically stimulated luminescence, a technique that offers a measure of the time passed since a mineral object has last seen the sunlight. This suggests that the ancestors of the First Nations people could have arrived in Australia much earlier, around 65,000 years ago or earlier (Figure 13).

Is this date for the arrival of Sapiens in Australia plausible if we accept, as the current prevailing scientific discovery dictates, that the first wave of Sapiens left Africa around 70,000 years ago? No, because it implies that the First Nations' Sapiens ancestors took only 5,000 years to cross about 12,000 miles (20,000 kilometers) of desert, mountains, jungles, and oceans. If, on the other hand, we consider the possibility that the first wave of Sapiens left Africa at least 135,000 years ago, then they would have had more than 70,000 years to accomplish the same journey.

A Demographic Progression

For these hunter-gatherers, we must assume that the march eastward was as unconscious as the exit from Africa. But still we must ask: what was the motive? The only conceivable explanation must be the tendency of Sapiens hordes to spawn new clans (with great collaborative skills and a division of tasks between people of both sexes). At around this time, 70,000 years ago, Sapiens would have first settled in all of

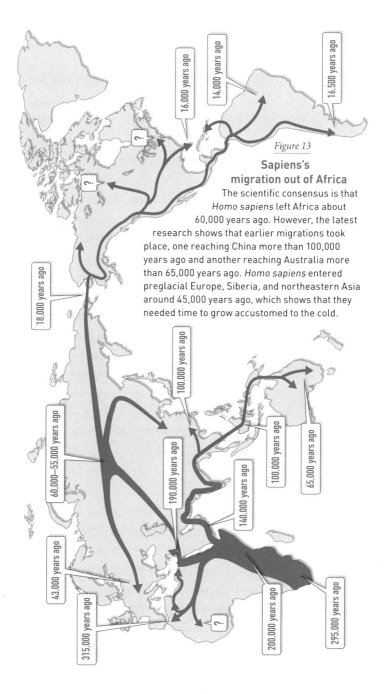

16,000 years ago

14,000 years ago

16,500 years ago

?

?

Figure 13

**Sapiens's
migration out of Africa**

The scientific consensus is that
Homo sapiens left Africa about
60,000 years ago. However, the latest
research shows that earlier migrations took
place, one reaching China more than 100,000
years ago and another reaching Australia more
than 65,000 years ago. *Homo sapiens* entered
preglacial Europe, Siberia, and northeastern Asia
around 45,000 years ago, which shows that they
needed time to grow accustomed to the cold.

18,000 years ago

100,000 years ago

60,000–55,000 years ago

190,000 years ago

140,000 years ago

100,000 years ago

65,000 years ago

43,000 years ago

315,000 years ago

200,000 years ago

?

295,000 years ago

southwestern Asia (southern Mesopotamia, and what is now Iran and Pakistan), followed by present-day India, Indonesia, and, finally, southern China.

Do we have evidence of an early Sapiens presence in the southern parts of Eurasia? Yes, because there is a series of dark-skinned populations of modern humans scattered between Africa and Australia. And what do we find when we examine the route as we've identified it? It's notable that southern Indians—Tamils in particular—are dark skinned. While the influence of the north is also perceptible (the result of later interbreeding with farmers from northern Eurasia), especially on the Indo-Chinese peninsula, the presence of dark-skinned populations is striking in the Andaman Islands (Indian territory close to Burma), the Indo-Chinese peninsula, the Philippines, and Malaysia. There are also the dark-skinned New Guineans and the First Nations people of Australia. This is an entire substratum of populations whose skin has remained dark because their ancestors never stopped living under the intense ultraviolet sunshine typical of the tropics. This is clearly a traceable path for the first wave of people migrating out of Africa.

On the other hand, we have very little fossil evidence along this route. But, of course, we also have little evidence from Africa at this time, and only a few Sapiens specimens. Surprising as this is, considering that the entire continent is where Sapiens finds its origin, it's easily explained by the fact that fossilization events are actually extremely rare in the tropics, because the heat activates the decay of flesh and bones. This is no different in India, south China, southeast Asia, and northern Australia.

With India specifically, questions about Sapiens's first appearance have increased with the discovery of the Jwalapuram site in the state of Andra Pradesh, in the southeast part of the country. Indian prehistorians have unearthed stone tools typical of the Middle Paleolithic period (300,000 to 30,000 years ago), distributed both above and below a layer of ash emitted by the mega-eruption of the Toba volcano on the island of Sumatra. This catastrophic event is firmly dated to 74,000 years ago, so we are given a hint—but a controversial one, since this site is the only one of its kind, and the presence of these stone tools is not accepted by all prehistorians—that Sapiens were in India prior to this eruption. But from when? Since there were already Sapiens in South China, we believe they lived in India too, as far back as 100,000 years ago, but this has not yet been proven archaeologically.

Let's consider the possibility, then, that some 40,000 years after clans of Sapiens migrated beyond Africa, they populated all of the Indian and Indochinese shores and perhaps part of the southeast Asian interior. Indeed, a 60,000-year-old Sapiens skull was found in Tam Pa Ling Cave in Laos, discovered by an international team that included including Fabrice Demeter of Paris's National Museum of Natural History.

The Very First Chinese Sapiens

We're almost certain that Sapiens arrived in tropical China with the first migration more than 100,000 years ago. In the Fuyan cave in Hunan (south China), a team led by Wu Liu of the Institute of Vertebrate Paleontology and

Paleoanthropology of the Chinese Academy of Sciences discovered forty-seven Sapiens teeth under a stalagmitic floor that was dated, using the uranium-thorium method, to around 80,000 years ago. These remains, like everything else found beneath this layer of calcite, are therefore older. The numerous animal bones that were also found there create a wildlife signature that confirms an age of more than 100,000 years. These very rare pieces of bone evidence suggest that the southern part of the Middle East, India, Indonesia, Australia, and South China all had populations of Sapiens much earlier than previously thought. If Australia was populated beginning around 65,000 years ago, and China more than 100,000 years ago, it's clear to us that the southern parts of Eurasia were colonized by Sapiens well before that.

Sapiens in the Cold

This explains the colonization of the warm regions of southern Eurasia, but what about northern Eurasia? As we've already discussed, the archaeological and ethnic evidence strongly suggest that the first waves of *Homo sapiens* left Africa more than 135,000 years ago, and then progressed through southern Eurasia without ever leaving the warm, tropical climates to which they were adapted. In both the Middle East and China, these first Sapiens from Africa, probably few in number, might have gone farther north and come in contact with non-Sapiens Eurasians, namely Neanderthals and Denisovans. This contact is especially evident in the Middle East, since Neanderthals had advanced to modern-day Iraq and the Levant some 100,000 years ago; but everything indicates that the humans there, the Neanderthals, were culturally very close to ancient *Homo sapiens*. Similarly, in central China, the first Eurasian Sapiens populations were able to mix with the local Eurasian populations, such as the Denisovans.

Thanks to recent advancements in biotechnology, we now have genetic evidence to back up this claim. First, the first complete sequencing of Denisovan genome—from a Denisovan who died in the cave of Denisova in central Siberia around 50,000 years ago—has taught us that crossbreeding between ancient Sapiens and local Neanderthals happened more than 100,000 years ago. Western Eurasia, where *Homo sapiens* met Neanderthals, provides clues as to how this crossbreeding occurred. In particular, while Sapiens had already moved into Australia, 12,000 miles (20,000 kilometers) away, they did not begin to successfully penetrate Europe—only 1,000 miles (1,500 kilometers) away—until around 43,000 years ago.

Sapiens the Hybrid

So how can such a shift be explained? Did Sapiens clash with Neanderthals? We don't believe this idea for a second. Prehistoric hunter-gatherer band societies (hordes), remember, were living in the wild; if they hadn't chosen hunting territories near other hordes, they might very well have never met and therefore disappeared due to genetic impoverishment. According to available paleodemographic studies, it has been estimated that the entire Neanderthal population never exceeded seventy thousand individuals. From the genomic diversity, paleogeneticists have estimated that of this population, only 10,000 were women of childbearing age. Taken together, this correlates to a very low population density, something on the order of .02 inhabitants per square mile (.01 inhabitants per square kilometer) within the huge Neanderthal territory.

After leaving Africa, *Homo sapiens* undoubtedly had an even lower population density. To explain their progress toward Australia, we must assume that in the first 40,000 years of their presence in Eurasia, they had to vigorously grow their population and that they must have greatly enhanced their effectiveness in the tropics, especially since they were already adapted to warm climates. Then they certainly would have proceeded to enrich themselves both genetically and culturally, by mixing their genes with those of non-Sapiens Eurasians. In Asia, these might have included some descendants of *Homo erectus*, and, thanks in part to paleogenetics, we now know that the Eurasian population before the arrival of Sapiens included mainly Neanderthals, Denisovans, and Flores Man (a fascinating ancient population of Lilliputian humans often referred to as "the hobbits," found on the island of Flores in Indonesia).

Neanderthal-Sapiens Crossbreeding

In terms of Neanderthal-Sapiens crossbreeding, we know that Sapiens encountered Levantine Neanderthals who were already acclimated to the warm weather of the Middle East. The culture and biology, however, of these Neanderthal hunter-gatherers from the north were better suited to the cold. So the cultural and biological interactions of Sapiens with

Denisovans and these Neanderthals could only prepare them to move north as well. It's likely that Sapiens advanced into Siberia and modern-day Russia before moving to Europe. A 45,000-year-old tibia, found in Ust'-Ishim in Siberia, is that of a Sapiens, whose ancestors had mixed with Neanderthals some 15,000 years earlier. And the DNA of a young man who died some 36,000 years ago at the Kostenki 14 site in Ukraine suggests an even earlier Neanderthal-Sapiens crossbreeding, making it even more likely that Sapiens ventured north around 60,000 years ago (Figure 14).

We must accept that we only have this data to write our history of the Sapiens's conquest of the north. It does suggest, however, that between 60,000 and 50,000 years ago, mixed populations and cultures had begun to form in the Middle East, Asia Minor, and probably farther into the interior of Asia, at the same latitude. Members of these populations, produced by the Sapiens-Neanderthal encounter, were likely to acquire survival techniques and the immune system necessary to live in the cold realm of Neanderthals.

As hybrid Sapiens, they were also likely to mix easily with newcomers from areas where Sapiens's population densities were higher. This suggests that, as in North America at the time of the European conquest, a sort of demographic coalition had formed, dominated by mixed populations and mixed cultures. Very slowly, over a period of approximately 20,000 years, there was a migration to the north, where the demographic dynamics of Sapiens gradually became dominant. Roughly 2,500 generations after the European disappearance of the last "typical Neanderthals," which were probably also

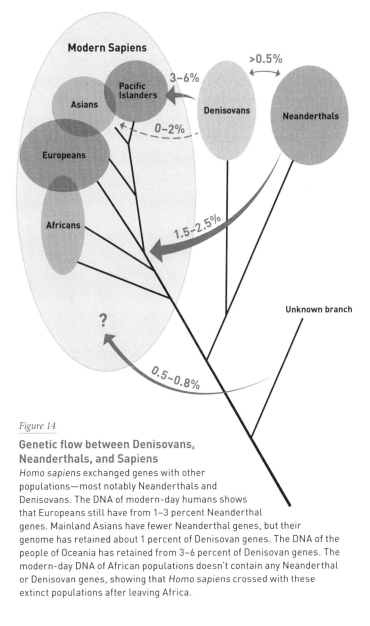

Modern Sapiens

Pacific Islanders

3–6%

>0.5%

Asians

Denisovans

Neanderthals

0–2%

Europeans

Africans

1.5–2.5%

?

Unknown branch

0.5–0.8%

Figure 14

Genetic flow between Denisovans, Neanderthals, and Sapiens

Homo sapiens exchanged genes with other populations—most notably Neanderthals and Denisovans. The DNA of modern-day humans shows that Europeans still have from 1–3 percent Neanderthal genes. Mainland Asians have fewer Neanderthal genes, but their genome has retained about 1 percent of Denisovan genes. The DNA of the people of Oceania has retained from 3–6 percent of Denisovan genes. The modern-day DNA of African populations doesn't contain any Neanderthal or Denisovan genes, showing that *Homo sapiens* crossed with these extinct populations after leaving Africa.

mixed with Sapiens, modern Eurasian humans have, on average, between 1 and 3 percent Neanderthal DNA. This is a considerable proportion, given the genetic erosion and many migrations that have taken place since then. It's a striking fact that many of these genes that have been preserved are those that deal with our adaptation to the cold or those that are involved in the formation of keratin (which makes up nails and hair and is responsible for skin pigmentation). Neanderthals' genes, it seems, facilitated *Homo sapiens*' adaptation to the Eurasian environment.

An overall picture emerges: after having more densely populated the south, from Africa to Australia, demographically dynamic Sapiens began to edge into Eurasia. Sapiens then met two populations, which were vaguely related thanks to the genetic flow across the Eurasian steppes: Neanderthals and Denisovans. The western Eurasians would therefore have been the product of crossbreeding between Sapiens from the western part of the southern climate and Neanderthals, the first westerners. Eastern Eurasians—notably the various Paleo-Asians—would have been the product of crossbreeding between the Sapiens who had reached the east through the steppes (and were carrying Neanderthals' genes) and Sapiens from the eastern part of the southern climate and the first far easterners, most likely the Denisovans.

Around 45,000 years ago, after this crossbreeding, traffic on the steppe highway would have enabled other exchanges, resulting in the Siberian, Amerindian, and Ainu (indigenous Japanese) populations. From south to north, across the planet, Sapiens had adapted to all types of climates and biomes and had profoundly modified both landscapes and animal populations, notably by causing the extinction of large mammalian predators, who were atop the terrestrial food chain after tens of millions of years of evolution. How can such incredible evolutionary success, if it is indeed success, be explained? Through culture and social structuring. Let's examine how.

CHAPTER 9

The Emergence of the Tribe

Early settlements of hordes, essentially tribes, or at least regional cultures, have existed since the "Aurignacian culture" of about 40,000 years ago in Europe. The Chauvet cave in Ardèche, France, indicates that these people quickly developed a social structure. The domestication of the wolf points to the emergence of more complex societies, as does that of humans, by humans. This phenomenon—self-domestication—was already in full swing, a fact that has been confirmed not only by our anatomy but also by evidence of enhanced technology, a semi-sedentary lifestyle, and the existence of social structure.

Forty thousand years ago, at the beginning of the Upper Paleolithic period (roughly 40,000 to 10,000 years ago), Sapiens had spread to conquer the ancient world. It would be another 20,000 years, when glaciation lowered sea levels by 400 feet (120 meters), before Sapiens reached the New World via the Bering Strait that separates modern-day Russia from Alaska. On every continent, human evolution was being driven almost entirely by social evolution. People were no longer living in small groups dependent on nature, but with an increasing frequency in much larger ordered groups, where the individual had to pay attention to social rules in order to get the resources necessary for survival, or even a chance to reproduce. This produced larger and more structured societies; the process that led to globalization had begun. This would be connected to an overall increase in the human presence in nature, which therefore became massively altered as Sapiens gradually moved from an economy of collection (with hunter-gatherers) to one of production (with stockpiling, farming, and livestock).

While people lived predominantly in hordes (band society), some 40,000 years ago at the beginning of the Upper Paleolithic period, today they live predominantly in large societies, some of which have billions of members. We will now explore various social systems that appeared in between these two extremes, which can be seen from archaeological site analysis. We will trace them primarily in Europe, because this is where we have the most prehistoric knowledge, and end with the advent of the first large-scale societies, state societies.

Social groups have a tendency to increase in size, and therefore the first form of society that appeared in the Upper Paleolithic was a constellation of hordes that shared a culture and became a tribe. In this sense, a tribe is a self-governing group of humans sharing a familial origin, real (biological) or imagined (cultural, social, or legendary attribution). We suggest that the tribe succeeded the horde at the beginning of the Upper Paleolithic because these social structures existed on every continent where Europeans set foot; contemporaneous writers observed and described the various types of tribes that were present in antiquity. The presence of tribes

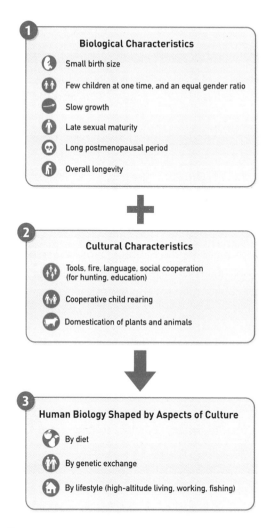

Figure 15

The "making" of Sapiens

Many features characterize *Homo sapiens*: anthropological traits typical of K-selection strategy (1), cultural traits that have helped to maintain and develop these biological features (2), and sociocultural traits that continue to influence our biology (3).

throughout the entire planet and the great variety of this social structure being surely the result of a long evolution, we
can infer that tribes may have existed very early in the Upper
Paleolithic.

Social Glue

These first tribes were likely created by population growth.
But even if several satellite hordes living near each other were
aware of a shared origin, a certain "social glue," a bond based
on shared interests or needs, would have been essential to
maintain a common culture. Ethnographic research suggests
that members of a hunter-gatherer horde would have worked
for only five hours each day and then shared what had been
collected. So why would these groups adopt the larger group
lifestyle? It would have been necessary, within a horde or
constellation, for some members to work actively to create a
shared culture that functioned as "social glue."

Their motives were probably diverse. They range from the
acquisition of social advantages to the creation of large teams
of hunters that, together, were able to take down large animals and thus obtain a great deal of resources at one time.
During the cold pre-glacial periods of the first half of the
Upper Paleolithic, the Eurasian steppes were home to innumerable herds of large herbivores, which would have been
easily spotted on the landscape, so their trapping would have
been easy to plan. A mammoth slaughtered 45,000 years
ago on the banks of the Yenisei River in central Siberia per-
fectly illustrates this phenomenon. The animal, which had
been stabbed many times by multiple hunters, was likely

deliberately corralled toward the muddy riverbank, where it became immobilized and would have been easily killed.

The Subsidized Artists of the Chauvet Cave

Whatever the exact personal or collective interests that led to the first conglomerations of hordes were, it is clear that some 40,000 years ago, socially structured tribes already existed and were probably fairly large. This is indicated by the Chauvet cave, in Ardèche, France, decorated with paintings made, for the most part, during the European cultural period known as the Aurignacian (43,000 to 28,000 years ago).

This cave, discovered in 1994 and declared a UNESCO World Heritage Site since 2014, is magnificently adorned with no fewer than 447 representations of animals, including 335 distinct drawings, the oldest of which dates to around 37,000 years ago. These animal representations showcase a mastery of shading and perspective—techniques that suggest movement—and composition, all of which would only be re-developed in Europe, indeed, not until antiquity or even the Renaissance. These drawings prove that in the Aurignacian era, in this part of Europe, there existed a society that could afford to train high-level specialists in the art of animal drawing. "Artists" of this level could only be the result of intense training, which would have required social organization. In his recent book *The Prehistory of Artistic Feeling*, the philosopher Emmanuel Guy points out that diverse styles and artistic personalities can be observed in the various works in the famous decorated caves of France and Spain. This implies that these animal representations were the works of professional artists (in a prehistoric sense), trained and supported by society. We deduce that within the Aurignacian culture, there were individuals influential enough to support a kind of "professional artistic life."

Does this "subsidized" activity in the Chauvet cave really constitute art? For the great prehistorian André Leroi-Gourhan (1911–86), who studied other painted caves such as the well-known Lascaux cave (dating from around 16,000 years ago), the animal drawings in the decorated caves were mythograms, abstract symbolic representations of myths. It's hard not to be struck by the atmosphere of animism (the belief that protective or hostile spirits *animate* living beings) or perhaps totemism (the extension of the notion of kinship to powerful animals, whose totemic clan possesses certain traits) that reigns in the cave. This suggests that Aurignacian society was composed of clans in which, as with more modern totemic cultures, shamans presided over communication with the clan's protective spirits. Dating from 40,000 years ago, an ivory sculpture of a mammoth discovered in a cave in Hohlenstein-Stadel in the Swabian Jura region of Germany seems to be a straightforward representation of a man-and-lion totem, perhaps a masked shaman in the process of overseeing some totemic rite. As you can see, early belief systems and rituals were most likely the principal focus of the founders of the first tribal cultures and also functioned as "social glue."

The Decorated Body

There's another indication of the existence of socially structured tribal societies during the Aurignacian period, in

Europe and elsewhere: body ornaments. Various environmental elements (pearls, animal teeth, shells, stones, and antlers) were used to adorn the body and garments. Ethnography has taught us that tribes—regional cultures in our present era—tend to cultivate their identities with certain markers, which helps explain the astonishing diversity of traditional textile costumes on the planet. In the Aurignacian period, this variety is most noticeable in ornamentation.

Thus, after having inventoried at least 162 different kinds of ornaments from 97 different habitats, the prehistorians Marian Vanhaeren and Francesco d'Errico of the CNRS found that the Aurignacians living in southeastern France, Italy, Austria, and the eastern Mediterranean used ornaments that were very different from those worn by the tribes living in Northern Europe. It's not, however, the presence or absence of raw materials in an area that explains the variety in adornment. Animal teeth used in southwest France, for example, came from a species also hunted in Italy, but the teeth were not transformed into ornaments there. It's obvious that regional cultural differences already existed, which corroborates the conclusion that tribal cultures existed as well.

Man's Wolf

The emergence of tribal societies coincides with the first domestication of an animal: the wolf. We don't know for certain if this major social innovation—the first instance of an animal in human society—occurred either in the Aurignacian period or the cultural period that followed, the Gravettian (31,000 to 22,000 years ago), first identified at the site of La Gravette, in France. According to genetic research published in 2017 by a team led by Bridgett M. vonHoldt of Princeton University, the wolf was domesticated in a single location between 40,000 and 20,000 years ago. This is supported by the discovery of presumed dog skulls in numerous caves, most notably in Razboinichya, in the Siberian Altai Mountains, and in Goyet, in Belgium. All these caves date to more than 30,000 years ago, before the Gravettian period in Western Europe.

The alterations to the genetic legacy of the wolf required to transform it into a dog focus on the reduction of skull size, modification to tail and ear shape, reduction in leg length, and reduction in fur length and density. But it goes even deeper than these changes, since humans also reduced the wolf's natural aggressivity in favor of more docile animals. In 2017, the study of twenty-nine specific genes known to play a role in the sociability of dogs indicated that the genes GTF2I and GTF2IRD1 could be the origins of their hypersociability, one of the keys to their coexistence with humans. Indeed, the alteration in these two genes is in part responsible for Williams syndrome, a rare genetic disorder, affecting many parts of the body, that causes humans to demonstrate a lack of social inhibition. Today, ethologists and biologists can identify and catalog, on both the anatomical and genetic level, the specific characteristics that are present in a domesticated species and absent in its wild ancestor. Taken together, they constitute what biologists call the *domestication syndrome*. With dogs, this syndrome is especially important and multifaceted, since, beginning in the Paleolithic period, humans have been constantly modifying dogs to adapt them to their needs. It's especially funny that, beginning in the Neolithic period, the agrarian era, the wolf's AMY2B gene was modified in a way that makes dogs able to digest bread.

Hunting With or Without One's Dog

We have given more jobs to the dog, this "wolf filled with humanity," in the magnificent words of the seventeenth-century French poet Jean de la Fontaine, than to any other animal. There are war dogs, shepherds, trackers, diagnosticians, companions, racing dogs, sled dogs, rescuers, personal aides, actors, attack dogs, retrievers, guard dogs, guide dogs, and much more. It can be assumed that, during the Paleolithic period, dogs had one major role: hunting assistant. The ethologist Pierre Jouventin of CNRS pointed out that hordes of wolves and men would have interacted frequently, since they occupied the same ecological niche. Wolves, however, can rarely kill a large animal, but they easily take advantage of the remains left by hunters. It is possible that many affinities have thus been created between hordes of wolves and men, who are undoubtedly at the origin of the domestication of the wolf, first by the modification of its genetic heritage by taming and then by coercion in training. Jouventin stresses that "the association between man and tamed wolves constituted a major adaptive advantage for humans. In fact, studies of Bushman lifestyle suggest that a hunter accompanied by a dog brings back three times more game."

Is Sapiens a Self-Domesticating Animal?

If we spend so much time outlining the domestication of

the wolf, it is because it signals and almost certainly dates the domestication of human beings . . . by human beings. As the ancients said, "A man is a wolf to another man" (*Homo homini lupus est*), and as we have said, humans welcomed the wolf into their society. They were able to do so because they were living in tribal societies that were much more co-ercive than hordes, since cohesion presumes all kinds of con-straints. What kinds, exactly? Working together, for example, and respecting various social statuses, observing rituals, not breaking certain taboos, sharing resources according to com-plicated rules, observing conjugal rules within a framework of kinship, etc. Originating in tribal thinking, the numerous rules for marriage give an example of this type of tribal con-trol on members' reproduction: in most religions, people are not supposed to marry outside their religious group. This is only one example of the tight control of the human group upon the individual. Every member of a tribe (and even more so in a more complex society) is, in a way, an animal that has been domesticated by the group.

The metaphor is attractive, but is it appropriate? The idea that humans domesticated themselves is not new. In 1868, in his book *The Variation of Animals and Plants under Domes-tication*, Charles Darwin explores this idea. As with animals, humans would have seen their aggressiveness decrease over time and their sociability increase. This "cultural" selection of human society, which biologists call "domestication syn-drome," exists in opposition to natural selection. Assisting the weakest members, for example, or forcing members to adopt behaviors and respecting a large number of rules become very real selective pressures. It was true of hordes, but even more so in tribal societies and subsequently in more complex societies. We have a great deal of proof of the self-domestication phenomenon, since the modern human bears the results of an obvious domestication syndrome: the size of its skull, which is 15 percent smaller than that of the Aurignacians (this is especially visible when the 28,000-year-old fossilized Cro-Magnon 1 *Homo sapiens* skull is compared to that of a modern human); its progressive gracilization, or reduction in bone mass; and its relatively small sexual dimorphism, when compared to other hominid species.

The intensification of this phenomenon is especially manifest in Gravettian culture, when large tribes appeared and exploited a vast territory between the Atlantic Ocean and the Ural Mountains. Their lifestyle relied on a great deal of mobility, which is evident in the proportions and thickness of their femurs and tibias, when compared to more sedentary species. It's possible, however, that they led semi-sedentary lives. While the Gravettians lived on the margins of expanding glaciers, they seem to have lived—at least during the harsh winters—in what the Czech prehistorian Jiří Svoboda considers more or less temporary "villages," or in neighboring caves located in valleys through which migratory animals passed. Mammoths were of particular interest to the Gravettians living on the steppes of Central Europe, where they specialized in hunting these large pachyderms in teams. They killed enough of them to build huts made of dozens of mammoth bones. This is why, since 2015, the American anthropologist Pat Shipman has been proposing that the Gravettian people, who hunted large animals such as mammoths, were the first to domesticate the wolf. The material culture left by Gravettians suggests that they lived together in societies that were bigger than those of the Aurignacians. They had developed a sewing needle with an eye, and, due to the cold, certainly knew how to sew very effective leather clothing. Equipped with straight spikes called "Gravettes," their hunting weapons were obviously designed to be used in conjunction with a propulsion system (a sling and a kind of bow, for example). These seem to have been very effective. All together, it's easy to envision Gravettians as boreal hunters similar to northern Native Americans.

Equally striking is the fact that they would have been able to gather in large hunting camps, such as Dolni Věstonice and Pavlov in Moravia or Krems-Wachtberg in Lower Austria.

These findings imply that Gravettians mastered many techniques for transport and storage—of seeds and grains, for example, but also of meat. Pemmican, a mixture of fat, dried meat, and berries prepared by the indigenous people of the American Plains, which can be consumed several years after its manufacture, most closely illustrates how meat would have been preserved. Among these techniques, basket making probably played a role; the discovery of cloth remnants on a kiln-fired figurine suggests that if they could make such finely interlaced material, the Gravettians could also produce other types of less refined interlaced materials, like baskets, rope, and mats. The discovery of millstones at several Gravettian sites (Bilancino II, Paglicci, Kostenki 16, Pavlov IV) indicates the harvest and transformation of grains and wild grasses. All of this, along with large-scale animal exploitation that provided a great number of resources, implies that Gravettians practiced an economy of collection, treatment, and storage.

As a result of a number of ethnographic studies, the French social anthropologist Alain Testart (1945–2013), in his now classic book *The Hunter-Gatherers, or the Origins of Inequality* (1982), and Canadian Brian Hayden, in his 2008 book *Man and Inequality*, showed that, while hunter-gatherers who practiced direct distribution (in which everything is shared immediately and openly) created societies with egalitarian tendencies, hunter-gatherers who practiced deferred distribution (in which everything is shared later on, according to social constraints) created unequal societies. Stocks had to be inventoried and guarded and their use regulated, which would have led to a kind of control, or even privatization. A society in which these powers are exercised is assumed to be more unequal than one in which everything is collective and open.

When the ecosystem is rich in resources but doesn't provide throughout the year, especially in winter—which was the case in the Mammoth Steppes—it's necessary to have social constraints, enforced, for example, by a chief, to guarantee that the group will dedicate itself to amassing a supply of resources when they're available, preserving meat, fish, and harvested plants (by butchering, pruning, grinding, drying, and smoking), and finally transporting them to a safe place under guard. The functioning of an economy this complex implies that social constraints abounded in Gravettian societies, suggesting that they were structured into different social groups.

A Hard Life for Women?

What do we know about the place of women in Upper Paleolithic societies? Not much, although we have a large number of women's figurines (some in fired clay) representing the same theme: steatopygia (containing large amounts of fat on the buttocks), with powerful breasts, an obvious vulva, and often—a very touching detail—adorned with a pregnancy belt. Should we see in these representations a kind of "sacralization" of motherhood or a very strong social pressure toward more maternity in a (semi-)sedentary context? This is a question that is very difficult to answer. What we can say is that the Gravettian sites in central Europe, such as those in Dolni Vestonice, Pavlov, show a diversification of the sectors of activity for leatherworking, manufacturing stone tools, bone, ivory, and sculptures, which is perhaps linked to social and gender division of tasks within the group.

One thing is certain: the study of some spectacular Gravettian graves (and the subsequent Epigravettian graves) shows that men and women were buried with the same flint-cut blades and adorned with the same ornaments, sometimes covered with ocher. We don't see any difference between male and female burials. Though these graves cannot inform us about the role of women in Gravettian society, they do confirm, without ambiguity, the existence of a social structure in Gravettian society.

The Tomb of a "Big Man"

The discovery at the site of Sungir, in Russia, near the city of Vladimir about 125 miles (200 kilometers) east of Moscow, unambiguously illustrates such a social structure. It's the 30,000-year-old tomb of a Gravettian "prince," or at least someone with enough influence in his society to have had a sizable workforce at his disposal. Having died of a throat injury at the age of forty-five, the Prince of Sungir was laid in his grave on his back, alongside the bodies of a young boy and girl, estimated to be around twelve and ten years old, respectively. The children were also on their backs, and arranged head to head, which seems to represent a ritual. All three of the deceased figures were covered with ocher.

The abundant funerary materials accompanying them are striking in their richness: in the case of the man, there was an ornate overcoat adorned with more than 2,900 mammoth-ivory pearls; a headdress decorated with seashells, squirrel tails, spears; and other offerings made of organic material that has not been preserved. The children were dressed in equally ornate clothing, adorned with about 5,000 pearls. Since these were one-third smaller than the Prince's pearls, prehistorians have concluded that the children's garments were made especially for this purpose. But their mere existence represents at least ten thousand hours of work (if we allot twenty minutes to make each mammoth-ivory pearl), not including the time needed to sew them onto the clothing. This suggests that either the children were sacrificed to accompany this influential man, or they were connected to him via a familial or social link and had been killed at the same time during a skirmish of some kind. In either case, such wealth and the commitment of so many hours of specialized labor prove that the three dead people were of a higher social class, for whom members of the Gravettian lower classes worked. It's undeniable that the self-domestication of Sapiens was already very advanced in Gravettian culture.

CHAPTER 10

War and the State

War likely appeared very early on in social life, probably before the last ice age, though we cannot prove it. Endemic in some areas as long as 15,000 years ago, war disrupts economies by introducing the effects of interhuman predation. The first signs of crop selection are perceptible before the last glaciation, and all of the components of agrarian life existed thousands of years in advance of the transition to agriculture and the domestication of livestock. The war economy continued to grow after the arrival of the farmers, and it would be the warriors who invented the state.

Impressive as it is, the tomb of the "Prince of Sungir" also illustrates the violence that existed in pre-glacial Gravettian society. The spectacular social stratification revealed in this grave, along with the obvious ostentation, must be interpreted as evidence of strong competition between great figures similar to the prince. Consider the injury that caused his death (a throat injury): ostentation was not always enough to guarantee victory in social competition. In a stockpiling society, a society where wealth existed, the temptation to have the upper hand on a competitor could imply a need for violence, leading to war. So there may have been wars in the Gravettian era—that is, according to our definition, coordinated and murderous attacks on one human group by another—but we do not have proof. In the Solutrean period (22,000 to 17,000 years ago), which followed the Gravettian in Europe during the period of greatest glaciation, temperatures plummeted to around −4° F (−20° C), so war was probably not a priority.

Warlike tendencies probably resumed with the Magdalenian reindeer hunters (17,000 to 12,000 years ago), the next cultural period in Western Europe, which produced the magnificently decorated caves in Lascaux (France) and Altamira (Spain). It's clear that these hunter-gatherers (and stockpilers) practiced cannibalism that has been interpreted by some as warlike. At Trou du Frontal, a cave in the province of Namur, the Belgian prehistorian Édouard-François Dupont (1841–1911) discovered a "vault filled with human bones" representing eighteen individuals, most of them women, who, it seems, had received the same "treatment as numerous nearby animals." He interpreted the site as the remains of a great ritual banquet on the occasion of the death of a "chief," whose family, he imagined, had also been sacrificed.

This macabre interpretation seems unconvincing in light of a similar scene at a cave in Gough, in Somerset, England. There the remains of a Magdalenian abattoir, roughly 15,000 years old, contain a mix of animal and human remains that bore all the traces of having been used for purposes of nutrition. One striking detail is the fact that three drinking cups had been made from pieces of human skulls, reminiscent of those discovered at the end of the nineteenth century, notably in Magdalenian caves in Placard, in Charente, France.

Are they a type of war trophy, like those Scythian or Gallic warriors would shape out of their enemy's cranium, or were they related to complex funerary practices?

In any case, if we want clear evidence of the emergence of war after the last ice age (18,000 years ago), we must leave Europe for Africa. A thousand years after the butchering of humans in the Gough cave, around 14,000 years ago, in what would become Upper Egypt, a group belonging to the local Qadan culture established a cemetery to bury members of its family in an orderly fashion, in a place called Jebel Saha-ba, which is now covered by Lake Nasser. Of the sixty-one skeletons of men, women, and children found among many other fragmented remains, at least 45 percent suffered violent deaths. Stone points had been stuck into the bones of twenty-one individuals, suggesting that they'd been attacked with spears or arrows, while other bones had cut marks on them. Some had wounds that had healed.

In 2016, twenty-seven skeletons found in Nataruk, Kenya, by a team led by Marta Mirazón Lahr of the University of Cambridge, provided additional evidence of warlike violence about 10,000 years ago, at the very end of the Paleolithic period. In addition to suffering numerous bone injuries, some of the people found in Nataruk had also been tied up before being killed. This and other details strongly suggest the existence of war between two groups of hunter-gatherers. The overall impression is that the societies to which the groups of Jebel Sahaba and Nataruk belonged had faced chronic conflict. At the end of the Paleolithic period, war had already become endemic.

The implications of the outbreak of permanent war in the evolution of humanity are immense, because it changes the natural environment in which tribes lived. Before war, nature would have been dangerous for an individual but rather safe for the group if members understood it well enough to avoid the dangers in their habitat. But after the advent of war, tribes were forced to share their environment with hostile competitors, which represented a serious mortal risk for the group as a whole and each of its members.

In an almost systematic way, tribal ethnography of the last four hundred years teaches us that the majority of tribes undertook and practiced endemic war. In his book *War Before Civilization*, the anthropologist Lawrence Keeley of the University of Illinois details the mortality rates, as estimated by ethnographers, of various tribes practicing endemic war. Among the Yanomami, a large tribal culture who live in the Amazon forest to this day, the number of male deaths represented up to 36 percent of the population, compared to 0.1 percent in industrialized Europeans during the twentieth century. The reality is that today's societies, despite how astonishingly violent they may seem to us, are much less so than those of the past.

Also worth noting is that in tribal societies of the past, attacking other tribes, which is obviously destructive in social terms, nonetheless became a way of life, to the point that the "warrior" figure became an enduring universal reference point. Like migration or major climatic events, war introduced new selective pressures to both individual humans and groups, which, we believe, must have had genetic consequences. In the aftermath of war, during which captives were taken, new groups would have formed as a result of these conquests, and certain genes would have multiplied, or, on the contrary, become rare to the point that they disappeared entirely. Genes were being intermingled in new ways as a result.

It goes without saying that since war has become endemic, there have always been enough "big men" to impose this activity on populations. For the elite class, war, on the one hand, is a quick way to obtain resources that they can distribute to advance in their society's ranks. It also spontaneously imposes a social coherence that's likely to reinforce their power. Examples collected by ethnologists show that in virtually all known tribal warrior cultures, prisoners of war were turned into slaves, meaning that war is probably the primary point of origin for the institution of slavery. Accordingly, leaders willingly started wars in order to enlarge their clans (with stolen women and children), to increase production (thanks to new workers), and to strengthen the clans' cohesion (by distributing spoils, including women, to their supporters). As a result of the competition between clans that made up tribes, this activity played an economic role in prehistoric tribal societies and, presumably, within prehistoric warrior societies.

Humanity's First Temples

The European Magdalanian period was followed in a more temperate Middle East by the Preneolithic (12,000 to 10,000 years ago), during which human groups became more sedentary as they became indisputable stockpilers (Figure 16). There is not yet any pottery from this era, but humans were already living in villages. Tribal leaders and religious elites received, from their groups, colossal contributions, in terms of effort and resources, to build temples, houses of worship, and monumental tombs. Examples like these are numerous in Turkey and Syria, in the Euphrates River Valley. The most impressive of these hunter-gatherer sanctuaries are those of the Gobekli Tepe (Turkey), built on a mountain: a series of anthropomorphic megaliths bearing engraved totems—each requiring months of work—were erected at regular intervals along the walls of an oval room lined with benches, and half buried, probably to support a roof.

The Neolithic Revolution That Never Happened

Let's examine one point relating to the evolution of Sapiens: one of archaeology's clichés, credited to the Australian Vere Gordon Childe (1892–1957), asserts that there was a "Neolithic revolution," a tremendous social leap during which hunter-gatherers, trying to escape their difficult hunting lives (a notion that has been debunked by studies of hunter-gatherer lifestyles), domesticated plants and animals, invented agriculture and husbandry, and settled down in communities to practice an economy based on production. Only at that point did they evolve a patriarchal, inegalitarian society that soon practiced continuous intertribal warfare.

Years ago

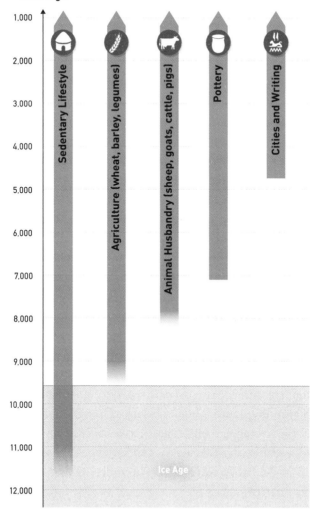

Source : J.-D. Vigne, 2008

Figure 16

The emergence of the Neolithic in the Middle East

The Neolithic economic system—i.e., pottery, animal domestication, and agriculture—developed slowly from the storage economy of the first sedentary hunter-gatherers. Pottery came after the cultivation of the first crops and the domestication of the first cattle.

But the Neolithic revolution never happened, or at least not in this way, since it was actually in the process of happening over a period of 20,000 years, with domestication (of the wolf and humans themselves) and the production economy (in the form of stockpiling), which appeared long before agriculture. The Gravettians had clearly already become semi-sedentary, domesticated an animal (the wolf), and practiced an economy of massive collection and storage, which points to a society already turning toward production.

The Ohalo II site, in Israel, which dates back 24,000 years, provides another striking example: on the shores of Lake Tiberias, a group of hunter-gatherers arrived and camped in huts, hunted, but primarily harvested cereal grains and other seeds, which they treated according to a very organized, step-by-step process, the first of which was storage. More than 90,000 wild cereal grain seeds of different kinds were found, along with traces of thirteen species of fruits and cereals. Seeds of "weeds" were mixed with emmer (wheat used as fodder), barley, and wild oats, whose subtle traits seem to prefigure the domesticated versions of these cereals. After also discovering flint blades bearing evidence of having been used to cut grasses (like wheat and barley), the team, led by Dani Nadel, an Israeli prehistorian, inferred that its occupants practiced, albeit on a small scale, an early form of grain culture more than 11,000 years before the famous beginnings of agriculture.

Thus, even though they lived before the last ice age, the people of Ohalo in the Middle East and the Gravettians in Europe were already semi-sedentary in the Preneolithic era.

Perhaps they hadn't yet entered the infernal cycle of war, but, as we've seen, the warlike conflicts seem to have already become widespread well before the Neolithic.

Beginning in the Neolithic period, war continued and continued and continued in most parts of the world, while peasants worked hard in the fields. Sedentarization, occurring at a greater level than previously observed, had a huge impact on human life (Figure 17). Studies of numerous bones taken from necropolises show that by encouraging the grouping of multitudes of people in one place, sedentarization created the ideal conditions for the transmission of infectious diseases. Many of them, such as tuberculosis, brucellosis, and measles, came from domesticated animals.

The agrarian way of life also reduced diversity in the diet. Having been broad consumers as hunter-gatherers, farmers became limited in their diet to the consumption of just a few cereals and animals. All of this made people more physiologically fragile in the face of infections, with overconsumption of certain compounds, such as slow-burning sugars in cereals. The new production economy, based on agriculture and livestock, also contributed to the dramatic population growth that we're still experiencing to this day.

Let's see where this increases comes from, and at what point it becomes worrying. The late paleodemographer Jean-Pierre Bocquet-Appel (1949–2018) of the CNRS attributed humanity's growth to cereal agriculture because it provided a richer, sweeter, and more accessible diet than would have been possible in hunter-gatherer cultures. The sedentary lifestyle also reduced the considerable stress that hunter-

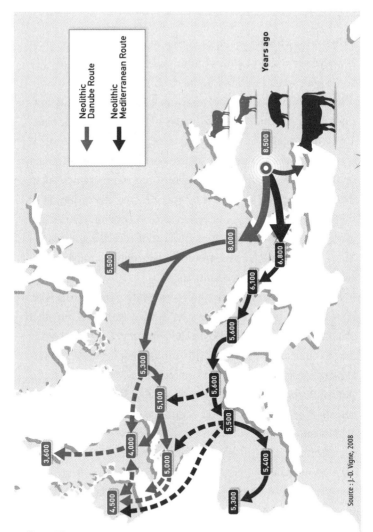

Figure 17

The introduction of domesticated animals to Europe
Neolithic culture appeared in many parts of the world, but it was brought to Europe from Asia Minor and the Levant, either by a Mediterranean route or through the Danubian corridor.

gatherers endured while moving around in the wild, with their children, in search of food resources.

In contemporary hunter-gatherer cultures, women have, on average, one child every three years, an interval attributed to the "energy balance," or the ratio between energy consumed through food and energy expended through breastfeeding and physical activity. But, while infant mortality was high in agrarian societies, this interval was reduced to just one year between children, since the human diet was regular and ensured. As a result, demographers estimate that, at the beginning of the Upper Paleolithic period (40,000 years ago), the world's population was less than 1 million—compared to almost 10 million by the end of this same period, about 10,000 years ago, and almost 100 million at the beginning of state civilizations, around 5,000 years ago. They predict that the world's population will reach 10 billion by 2050 (Figure 18).

From the Neolithic to the State

As Jean Guilaine of the Collège de France wrote in his 2011 book *Cain, Abel, Ötzi*, humans continue to practice the Neolithic way of life. This was especially true during the Copper Age (beginning 5,000 years ago), as the tragic fate of Ötzi illustrates. Ötzi, also called the "Iceman," is a well-mummified man that was found in September 1991 in the Ötztal Alps, near the border between Austria and Italy. The study of the artifacts and clothes of this man suggested that Öetzi was a notable member of a small alpine tribe, who was killed some 5,300 years ago by an arrow in his scapula while he was trying to escape across a glacier. Elsewhere in Europe, so-called big men

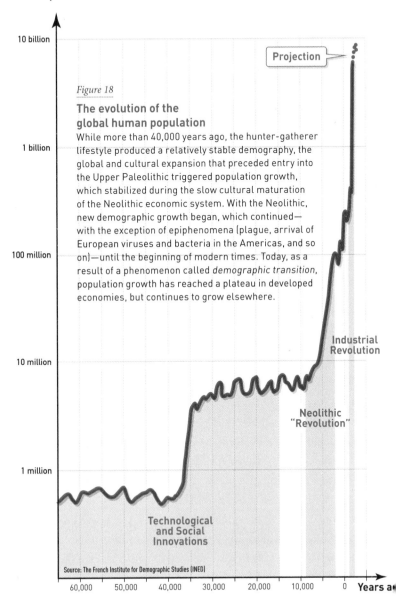

Population

Figure 18

The evolution of the global human population

While more than 40,000 years ago, the hunter-gatherer lifestyle produced a relatively stable demography, the global and cultural expansion that preceded entry into the Upper Paleolithic triggered population growth, which stabilized during the slow cultural maturation of the Neolithic economic system. With the Neolithic, new demographic growth began, which continued—with the exception of epiphenomena (plague, arrival of European viruses and bacteria in the Americas, and so on)—until the beginning of modern times. Today, as a result of a phenomenon called *demographic transition*, population growth has reached a plateau in developed economies, but continues to grow elsewhere.

Projection

10 billion

1 billion

100 million

10 million

Industrial Revolution

Neolithic "Revolution"

1 million

Technological and Social Innovations

Source: The French Institute for Demographic Studies (INED)

60,000   50,000   40,000   30,000   20,000   10,000   0   **Years a**

had already become great warriors within tribes whose members sometimes numbered into the thousands. It is among tribes like these that, in Upper Egypt, Mesopotamia, the Indus Valley, and what would become China, warlords surrounded by multitudes of warriors would eventually arise.

Forebears of elite troops that have always surrounded kings, these warriors and other bodyguards were fanatically devoted to their leader for one simple reason: in many cultures, they would be put to death as a matter of principle if he died. Therefore, losing or failing on the battlefield was unimaginable. For Alain Testart, some of these fanatic warriors became the essential lever by which certain leaders imposed their rule on groups that had become too large for these chiefs to know all their members. Once their domination had been assured through war, they had to become administrators. And thus the first states were born, and humanity moved one step closer to globalization.

Conclusion

Does Sapiens's past tell us anything useful about its future? Since the creation of the first proto-states in Mesopotamia and Egypt, as early as the fourth millennium before our own era, societies have continued to grow despite innumerable upheavals. While there were fewer than 1 million Sapiens on the surface of Earth at the beginning of the Upper Paleolithic, there were almost ten million at the end, almost one hundred million at the beginning of the metallurgical ages, and almost 1 billion by the year 1800, at the beginning of the industrial era. We are now at around 7.5 billion.

All of this is clearly frightening: too populous, humanity devastates nature. It is estimated that about 1 billion of us live—poorly—in slums, and our atmosphere is warming very quickly. The impact of human life on the rest of the planet is so strong that geologists have named the current geological era the Anthropocene, the "human era."

Sapiens, however, will continue to evolve. More than ever, we are social animals whose environment is no longer the wilderness but human society. For many millions of years, our evolution was more biological than cultural. And then, a few hundred thousand years ago, *Homo sapiens* appeared, and its evolution has been more cultural than biological, reaching a point some 40,000 years ago when culture overtook biology.

Very quickly, we went from gatherers to producers of food, becoming a species that always produces more to reproduce more. This pattern explains how we're now approaching a population of 10 billion humans, which, considering the current condition of our social organization, is much too much for our planet, and especially for life on Earth. The last form of collection in nature—fishing—is threatened with extinction by overexploitation; food is industrial and often comes from plants with deliberately altered genes; these genetically modified organisms are fed to domestic animals, which are transformed into feeding machines; robots have taken the place of warriors; we have more contact with humans that we will never see in person than with those we will; and social complexity has become so great that it irreversibly exceeds the human mind, even that of our leaders.

We are approaching the day when we consume more in a year than the planet can produce, while entire ecosystems disappear. It is clear, however, that a new anthropological transition is beginning that will take us out of the Neolithic mentality toward a new type of psyche. But what kind, exactly? It's still difficult to know, but what's happening in developed countries offers some clues: people are having fewer children, and a lot of what is produced is now done in a virtual space. The pleasure of living and the need to make sense of life are important values in today's culture. In developed countries, digital technologies have changed our lifestyles to allow for more freedom and opportunity than ever before. Many modern-day migrations involve people from developing countries moving to countries where it's almost as if all communication and every action is subtly and secretly prepared by an invisible army of robots, suggesting that this transition is accelerating.

Since we now possess a sort of global nervous system—the internet—humanity is changing. Almost everyone can now communicate with each other. A new digital cultural lifeblood is flowing increasingly faster, through half—or more—of humanity. There is no doubt it will continue to spread more and more, to change life on Earth. But we want to stay optimistic. Even though it might not seem very obvious, Sapiens remains sapiens, which is to say "wise." And we'd wager that, over time, we will become even wiser.

References

Chapter 1: A Biped Descends from an Ape

M. Brunet et al. "A new hominid from the Upper Miocene of Chad, Central Africa." *Nature*, 418 (2002): 145–51.

B. Senut et al. "First hominid from the Miocene (Lukeino Formation, Kenya)." *Comptes Rendus de l'Académie de Sciences*, 332 (2001): 137–44.

M. Pickford et al. "Bipedalism in *Orrorin tugenensis* revealed by its femora." *Comptes Rendus de l'Académie de Sciences,* 228, no. 4 (2002): 191–203.

P. G. M. Dirks et al. "Geological setting and age of Australopithecus sediba from Southern Africa." *Science*, 328 (2010): 205–8.

T. D. White et al. "*Ardipithecus ramidus* and the Paleobiology of Early Hominids." *Science*, 326 (2009): 64–86.

Y. Haile-Selassie, G. Suwa, and T. D. White. "Late Miocene Teeth from Middle Awash, Ethiopia, and Early Hominid Dental Evolution." *Science* 303, no. 5663 (2004): 1503–05.

C. V. Ward et al. "Complete Fourth Metatarsal and Arches in the Foot of *Australopithecus afarensis*." *Science*, 331 (2011): 750–53.

M. D. Leakey. "The Fossil Footprints of Laetoli," *Scientific American*, 246 (1982): 50–57.

J. Kappelman et al. "Perimortem fractures in Lucy suggest mortality from fall out of tall tree." *Nature*, 537 (2016): 503–7.

Y. Deloison. "Etude des restes fossiles des pieds des premiers hominidés: *Australopithecus* et *Homo habilis*. Essai d'interprétation de leur mode de locomotion. Ph.D. thesis under the direction of Yves Coppens," Université de Paris V Sorbonne, 1993.

S. Bortolamiol et al. "Suitable habitats for endangered frugivorous mammals: small-scale comparison, regeneration forest and chimpanzee density in Kibale National Park, Uganda," *PLoS ONE*, 9 (2014): e102177.

Y. Coppens. *Le Singe, l'Afrique et l'homme*. Paris: Fayard, 1983.

D. C. Johanson. *Lucy's Legacy: The Quest for Human Origins*. New York: Harmony Books, 2009.

I. Tattersall. *Masters of the Planet: The Search for Our Human Origins*. New York: St. Martin's Griffin, 2013.

I. Tattersall and R. DeSalle. *The Accidental Homo Sapiens: Genetics, Behavior, and Free Will*. New York: Pegasus, 2019.

Chapter 2: Culture, the Evolutionary Accelerator

S. Harmand et al. "3.3-million-year-old stone tools from Lomekwi 3, West Turkana, Kenya." *Nature*, 521 (2015): 310–15.

J. Goodall. *My Friends, the Wild Chimpanzees*. Washington, DC: National Geographic Society, 1967.

B. Villmoare et al. "Early Homo at 2.8 Ma from Ledi-Geraru, Afar, Ethiopia." *Science*, 347, (2015), 1352–55.

H. Roche. "Cognition et Culture matérielle." Talk to the Fyssen Foundation, when she received the Prix Fyssen 2012.

J. C. Thompson et al. "Taphonomy of fossils from the hominin-bearing deposits at Dikika, Ethiopia." *Journal of Human Evolution*, 86 (2015): 112–35.

L. C. Aiello and R. I. M. Dunbar. "Neocortex Size, Group Size, and the Evolution of Language." *Current Anthropology*, 34, no. 2 (1993): 184–93.

L. S. B. Leakey et al. "A New Species of The Genus *Homo* From Olduvai Gorge." *Nature*, 202 (1964): 7–9.

B. A. Wood. *Koobi Fora Research Project Volume 4: Hominid Cranial Remains*. New York: Oxford University Press, 1991.

B. A. Wood and M. Collard. "The Human Genus." *Science*, 284 (1999): 65–71.

R. Klein, *The Human Career: Human Biological and Cultural Origins*, 3rd ed. Chicago: University of Chicago Press, 2009.

Chapter 3: My Big Head (Almost) Killed Me

J. Giedd et al. "Brain Development during Childhood and Adolescence: A Longitudinal MRI Study." *Nature Neuroscience*, 2 (1999): 861–63.

T. D. Weaver and J. J. Hublin. "Neandertal birth canal shape and the evolution of human childbirth." Proceedings of the National Academy of Sciences, 106, no. 20 (2009): 8151–56.

N. Roach et al. "Elastic energy storage in the shoulder and the evolution of high-speed throwing in *Homo*." *Nature*, 498 (2013): 483–86.

R. Caspari and S. H. Lee. "Is human longevity a consequence of cultural change or modern biology?" *American Journal of Physical Anthropology*, 129, no. 4 (2006): 512–17.

R. Caspari and S. H. Lee. "Older age becomes common late in human evolution." Proceedings of the National Academy of Sciences, 101, no. 30 (2004): 10895–900.

H. Pontzer et al. "Metabolic acceleration and the evolution of human brain size and life history." *Nature*, 533 (2016): 390–92.

L. C. Aiello and P. Wheeler. "The Expensive-Tissue Hypothesis: The Brain and the Digestive System in Human and Primate Evolution." *Current Anthropology*, 36, no. 2 (1995): 199–221.

K. Fonseca-Azevedo and S.Herculano-Houzel. "Metabolic constraint imposes tradeoff between body size and number of brain neurons in human evolution." Proceedings of the National Academy of Sciences, 109, no. 45 (2012): 18571–76.

A. Gibbons. "Food for Thought." *Science*, 316 (2007): 1558–60.

A. Navarrete et al. "Energetics and the evolution of human brain size." *Nature*, 480 (2011): 91–93.

C. K. Brain and A. Sillent. "Evidence from the Swartkrans cave for the earliest use of fire." *Nature*, 336 (1988): 464–66.

R. Wrangham. *Catching Fire: How Cooking Made Us Human*. New York: Basic Books, 2009.

N. Alperson-Afil and N. Goren-Inbar "Out of Africa and into Eurasia with controlled use of fire: Evidence from Gesher Benot Ya'aqov, Israel." *Archaeology, Ethnology and Anthropology of Eurasia*, 28, no. 1 (2006): 63–78.

Chapter 4: What Obligate Bipedalism Has Made Us

S. Semaw et al. "2.5-million-year-old stone tools from Gona, Ethiopia." *Nature*, 385 (1997): 333–36.

D. E. Lieberman. "Four legs good, two legs fortuitous: Brains, brawn and the evolution of human bipedalism." In *In the Light of Evolution: Essays in the Laboratory and Field*. J. B. Losos, ed. Greenwood Village, CO: Roberts and Company, 2011, 55–71.

D. E. Lieberman. *The Story of the Human Body: Evolution, Health, and Disease*. New York: Pantheon, 2013.

J. du Chazaud. *Les Glandes Endocrines : Leur rôle sur la sexualité et le système nerveux*. Paris: Éditions du Dauphin, 2016.

N. G. Jablonski and G. Chaplin. "The evolution of human skin coloration." *Journal of Human Evolution*, 39 (2000): 57–106.

N. G. Jablonski. *Skin: A Natural History.* Berkeley: University of California Press, 2006.

Chapter 5: Hunting Arouses All of the Senses

A. Stoessel et al. "Morphology and function of Neandertal and modern human ear ossicles." *Proceedings of the National Academy of Sciences*, 2016.

L. Werdelin and M. E. Lewis. "Temporal Change in Functional Richness and Evenness in the Eastern African Plio-Pleistocene Carnivoran Guild." *PLoS ONE*, 8, no. 3 (2013): 1–11.

K. G. Hatala et al. "Footprints reveal direct evidence of group behavior and locomotion in *Homo erectus*." *Scientific Reports*, 6 (2016).

M. R. Bennett et al. "Preserving the Impossible: Conservation of Soft-Sediment Hominin Footprint Sites and Strategies for Three-Dimensional Digital Data Capture." *PLoS ONE*, 8, no. 4 (2013): e60755.

J. M. Harris, ed. *The Fossil Ungulates: Geology, Fossil Artiodactyls and Palaeoenvironments*. Koobi Fora Research Project, 3, no. 1, New York: Oxford University Press, 1991.

H. L. Dingwall et al. "Hominin stature, body mass, and walking speed estimates based on 1.5 million-year-old fossil footprints at Ileret, Kenya." *Journal of Human Evolution*, 64 no. 6, (2013): 556–68.

C. Hobaiter and R. W. Byrne. "Flexibilité et intentionnalité dans la communication gestuelle chez les grands singes." *Revue de Primatologie*, 5 (2014).

C. Crockford et al. "Wild Chimpanzees Inform Ignorant Group Members of Danger." *Current Biology*, 22, no. 2 (2012): 142–46.

R. D'Anastasio et al. "Micro-Biomechanics of the Kebara 2 Hyoid and Its Implications for Speech in Neanderthals." *PLoS ONE*, 8, no. 12 (2013): e82261.

J. A. Hurst et al. "An extended family with a dominantly inherited speech disorder." *Developmental Medicine & Child Neurology*, 32, no. 4 (1990): 352–55.

J. Krause et al. "The Derived FOXP2 Variant of Modern Humans Was Shared with Neandertals." Current Biology, 17, no. 21 (2007): 1908–12.

A. Flinker et al. "Redefining the role of Broca's area in speech." *Proceedings of the National Academy of Sciences*, 112, no. 9 (2015): 2871–75.

Chapter 6: The First Conquest of the Planet

L. Gabunia and A. Vekua. "A Plio-Pleistocene hominid from Dmanisi, East Georgia, Caucasus." *Nature*, 373 (1995): 509–12; L. Gabunia et al. "Earliest Pleistocene Hominid Cranial Remains from Dmanisi, Republic of Georgia: Taxonomy, Geological Setting, and Age." *Science*, 288 (2000): 1019–25.

M. Sahnouni et al. "1.9-million- and 2.4-million-year-old artifacts and stone tool–cutmarked bones from Ain Boucherit, Algeria." *Science* 362, no. 6420 (2018): 1297–1301.

M. J. Morwood et al. "Revised age for Mojokerto 1, an early *Homo erectus* cranium from East Java, Indonesia." *Journal Australian Archaeology*, 57, no. 1 (2003): 1–4.

M. Rasse et al. "The site of Longgupo in his geological and geomorphological environment." *L'Anthropologie*, 115, no. 1 (2011): 23–39.

R. Ciochon and R. Larick. "Early *Homo erectus* Tools in China." *Archaeology*, 53, no. 1, (2000): 14–15.

H. Alçiçek and M. Alçiçek. "Geographic and geological context of the Kocabaş site, Denizli Basin, Anatolia, Turkey." *L'Anthropologie*, 118, no. 1 (2014): 11–15.

H. de Lumley. "Le site de l'Homme de Yunxian, province du Hubei, Chine. Signification du matériel archéologique et paléontologique sur le site de l'Homme de Yunxian (note d'information)." *Comptes rendus des séances de l'Académie des Inscriptions et Belles-Lettres*, 153, no. 1 (2009): 119–23.

M. Pavia et al. "Stratigraphical and palaeontological data from the Early Pleistocene Pirro 10 site of Pirro Nord (Puglia, south eastern Italy)." *Quaternary International*, 2012.

I. Toro-Moyano et al. "L'industrie lithique des gisements du Pléistocène inférieur de Barranco León et Fuente Nueva 3 à Orce, Grenade, Espagne." *L'Anthropologie*, 113, no. 1 (2009): 111–24.

A. E. Lebatard et al. *Earth and Planetary Science*, 390 (2004): 8–18.

C. J. Lepre et al., "An earlier origin for the Acheulian," *Nature*, 477 (2011): 82–85.

I. Toro-Moyano et al. "The oldest human fossil in Europe dated to ca. 1.4 Ma at Orce (Spain)." *Journal of Human Evolution*, 65 (2013): 1–9.

O. Bar-Yosef et al. *The Lithic Assemblages of 'Ubeidiya, a Lower Paleolithic Site in the Jordan Valley*. Qedem 34: Publication of the Institute of Archeology, The Hebrew University of Jerusalem, 1993.

N. Ashton et al. "Hominin Footprints from Early Pleistocene Deposits at Happisburgh, UK." *PLoS ONE*, 9, no. 2 (2014): e88329.

J. Krause et al. "The complete mitochondrial DNA genome of an unknown hominin from southern Siberia." *Nature*, 464, no. 7290 (2010): 894–97.

Chapter 7: And *Homo sapiens* Emerged . . .

I. Hershkovitz et al. "The Earliest Modern Humans outside Africa." *Nature*, 520, no. 7546 (2015): 216–19.

J. J. Hublin et al. "New fossils from Jebel Irhoud, Morocco and the pan-African origin of *Homo sapiens*." *Nature*, 546, no. 7657 (2017): 289–92; D. Richter et al. "The age of the hominin fossils from Jebel Irhoud, Morocco, and the origins of the Middle Stone Age." *Nature*, 546, no. 7657 (2017): 293–96.

E. Scerri et al. "Did Our Species Evolve in Subdivided Populations across Africa, and Why Does it Matter?" *Trends in Ecology and Evolution*, 33, no. 8 (2018): 582–94; H. S. Groucutt et al. "*Homo sapiens* in Arabia by 85,000 years ago." *Nature Ecology & Evolution*, 2, no. 5 (2018): 800–9.

Y. N. Harari. *Sapiens: A Brief History of Humankind*. New York: Harper, 2015.

S. Condemi and F. Savatier. *Néandertal, mon frère*. Paris: Flammarion, 2016.

S. Sankararaman et al. "The genomic landscape of Neanderthal ancestry in present-day humans." *Nature*, 507, no. 7492 (2014): 354–57.

B. Vernot and J. M. Akey. "Resurrecting Surviving Neandertal Lineages from Modern Human Genomes." *Science*, 343, no. 6174 (2014): 1017–21.

A. Rosas et al. "The growth pattern of Neandertals, reconstructed from a juvenile skeleton from El Sidrón (Spain)." *Science*, 357, no. 6357 (2017): 1282–87.

P. Gunz et al. "Brain development after birth differs between Neanderthals and modern humans." *Current Biology*, 20, no. 21 (2010): 921–22; E. Pearce et al. "New insights into differences in brain organization between Neanderthals and anatomically modern humans." *Proceedings of the Royal Society of Biological Sciences*, 280, no, 1758 (2013); T. Kochiyama et al. "Reconstructing the Neanderthal brain using computational anatomy." *Nature*, 8 (2018): 6296.

E. Trinkaus and M. R. Zimmerman. "Trauma among the Shanidar Neandertals." *American Journal of Physical Anthropology*, 57, no. 1 (1982): 61–76.

F. de Waal, *Mama's Last Hug: Animal Emotions and What They Tell Us about Ourselves*. New York: W. W. Norton, 2019.

J. L. Arsuaga et al. "Sima de los Huesos (Sierra de Atapuerca, Spain). The site." *Journal of Human Evolution*, 33 (1997): 109–27.

T. Higham et al. "The timing and spatiotemporal patterning of Neanderthal disappearance." *Nature*, 512 (2014): 306–9.

S. G. Shamay-Tsoory. "The neural bases for empathy." *Neuroscientist*, 17, no. 1 (2011): 18–24.

Q. Fu et al. "The genetic history of Ice Age Europe." *Nature*, 534 (2016): 200–5.

Q. Fu et al. "An early modern human from Romania with a recent Neanderthal ancestor." *Nature*, 524 (2015): 216–19; R. E. Green et al. "A complete Neanderthal mitochondrial genome sequence determined by high-throughput sequencing." *Cell*, 134 (2008): 416–26.

R. E. Green et al. "A draft sequence of the Neandertal genome." *Science*, 328 (2010): 710–22.

Chapter 8: The Spread of *Homo sapiens* over the Entire Planet

K. Harvati et al., Apidima Cave fossils provide earliest evidence of *Homo sapiens* in Eurasia, *Nature*, 571 (2019): 500–4.

S. J. Armitage et al. "The Southern Route 'Out of Africa': Evidence for an Early Expansion of Modern Humans into Arabia." *Science*, 331 (2011): 453–56; A. Lawler. "Did Modern Humans Travel out of Africa via Arabia?" *Science*, 331 (2011): 387.

C. Clarkson et al. "Human occupation of northern Australia by 65,000 years ago." *Nature*, 547 (2017): 306–10.

L. Wu et al. "Human remains from Zhirendong, South China, and modern human emergence in East Asia." *Proceedings of the National Academy of Sciences*, 107, no. 45 (2010): 19201–6; X. Song et al. "Hominin Teeth From the Early Late Pleistocene Site of Xujiayao, Northern China," *American Journal of Physical Anthropology*, 156 (2015): 224–40.

R. E. Frisch. "The right weight: body fat, menarche and ovulation." *Baillière's Clinical Obstetrics and Gynaecology*, 4, no. 3 (1990): 419–39.

L. R. Botigué et al. "Ancient European dog genomes reveal continuity since the Early Neolithic." *Nature Communications*, 8 (2017): e16082.

X. Song et al. "Hominin Teeth from the Early Late Pleistocene Site of Xujiayao, Northern China," op. cit.

C. Hill et al. "Phylogeography and Ethnogenesis of Aboriginal Southeast Asians." *Molecular Biology and Evolution*, 23, no. 12 (2006): 2480–91.

R. E. Green et al. "A draft sequence of the Neandertal genome." *Science*, 328 (2010): 710–22.

A. Seguin-Orlando et al. "Genomic structure in Europeans dating back at least 36,200 years," *Science*, 346, no. 6213 (2014): 1113–18.

B. Vandermeersch. *Les Hommes fossiles de Qafzeh (Israël)*. Paris: Les éditions du CNRS, 1981.

M. Kuhlwilm et al. "Ancient gene flow from early modern humans into Eastern Neanderthals." *Nature*, 530 (2016): 429–33.

Chapter 9: The Emergence of the Tribe

S. L. Kuhn and M. C. Stiner. "What's a Mother to Do? The Division of Labor among Neandertals and Modern Humans in Eurasia." *Current Anthropology*, 47, no. 6 (2006): 953–80.

E. Guy. *Ce que l'art préhistorique dit de nos origines*. Paris: Flammarion, 2017.

A. Leroi-Gourhan. *Le Geste et la Parole: Technique et langage* (vol. 1); *Mémoire et les Rythmes* (vol. 2). Paris: Albin Michel, 1964–1965.

J.-P. Bocquet-Appel and A. Degioanni. "Neanderthal Demographic Estimates." *Current Anthropology*, 54, no. 58 (2013): 202–13; S. L. Kuhn and E. Hovers. "Alternative Pathways to Complexity: Evolutionary Trajectories in the Middle Paleolithic and Middle Stone Age," op. cit.

A. Testart. *Les chasseurs cueilleurs ou l'origine des inégalités*. Paris: Société d'Ethnographie, 1982.

B. Hayden. *L' homme et l'inégalité: l'invention de la hiérarchie à la préhistoire*. Paris: CNRS Editions, 2008.

C. Darwin. *The Expression of the Emotions in Man and Animals*. London: John Murray, 1871 (reprinted 1998, Oxford University Press).

C. Darwin. *The Variation of Animals and Plants under Domestication*, vol. II. New York: D. Appleton and Company, 1899.

A. S. Wilkins et al. "The 'Domestication Syndrome' in Mammals: A Unified Explanation Based on Neural Crest Cell Behavior and Genetics." *Genetics*, 197, no. 3 (2014): 795–808.

P. Jouventin. *Trois prédateurs dans un salon: Une histoire du chat, du chien et de l'homme*. Paris: Éditions Belin, 2014.

P. Shipman. *The Invaders: How Humans and Their Dogs Drove Neanderthals to Extinction*. Cambridge, MA: Harvard University Press, 2015.

A. Revedin et al. "Thirty thousand-year-old evidence of plant food processing." *Proceedings of the National Academy of Sciences*, 10, no. 44 (2010): 18815–19.

J. K. Kozłowski "The origin of the Gravettian." *Quaternary International*, 359–60 (2015): 3–18.

J. M. Adovasio et al. "Perishable Industries from Dolní Vestonice I: New Insights into the Nature and Origin of the Gravettian." *Archaeology, Ethnology and Anthropology of Eurasia*, no. 6 (2001): 48–65.

O. Soffer et al. "Recovering Perishable Technologies through Use Wear on Tools: Preliminary Evidence for Upper Paleolithic Weaving and Net Making." *Current Anthropology*, 45, no. 3 (2004): 407–13.

Chapter 10: War and the State

É. Dupont. *Les Temps Préhistoriques en Belgique: L'Homme pendant les Âges de la Pierre dans Les Environs de Dinant-sur-Meuse*. Bruxelles: C. Muquardt, 1873 (reprinted 2010, Kessinger Legacy Reprints).

R. Robert and A. Glory. "Le culte des crânes humains aux époques préhistoriques." *Bulletins et Mémoires de la Société d'Anthropologie de Paris*, 8, no. 1 (1947): 114–33.

F. Le Mort and D. Gambier. "Cutmarks and Breakage on the Human Bones from Le Placard (France)." *Anthropologie*, 29, no. 3 (1991): 189–94.

S. Bello et al. "Earliest Directly-Dated Human Skull-Cups." *PLoS ONE*, 6, 2 (2011): e17026.

D. Antoine et al. "Revisiting Jebel Sahaba: New Apatite Radiocarbon Dates for One of the Nile Valley's Earliest Cemeteries." *American Journal of Physical Anthropology Supplement*, 56 (2013).

R. Kelly. "The evolution of lethal intergroup violence." *Proceedings of the National Academy of Sciences*, 102 (2005): 24–29.

M. Judd. "Jebel Sahaba Revisited." *Archaeology of Early Northeastern Africa, Studies in African Archaeology*, 9 (2006): 153–66.

M. Lahr et al. "Inter-group violence among early Holocene hunter-gatherers of West Turkana, Kenya." *Nature*, 529, no. 7586 (2016): 394–98.

L. H. Keeley. *War before Civilization: The Myth of the Peaceful Savage*. New York: Oxford University Press, 1996.

G. Magli. "Sirius and the project of the megalithic enclosures at Gobekli Tepe." *Nexus Network Journal*, 18, no. 2 (2016): 337–46.

V. G. Childe. *Man Makes Himself*. London: Watts & Co., 1936 (reprinted 1951, New American Library).

D. R. Piperno et al. "Processing of wild cereal grains in the Upper Palaeolithic revealed by starch grain analysis." *Nature*, 430 (2004): 670–73.

D. Nadel et al. "Stone Age Hut in Israel Yields World's Oldest Evidence of Bedding. *Proceedings of the National Academy of Sciences*, 101 (2004): 6821–6826.

A. Snir et al. "The Origin of Cultivation and Proto-Weeds, Long Before Neolithic Farming." *PLoS ONE*, 10, vol. 7 (2015): e0131422.

J.-P. Bocquet-Appel and O. Bar-Yosef, eds. *The Neolithic Demographic Transition and its Consequences*. New York: Springer, 2008.

J. Guilaine. *Caïn, Abel, Ötzi: L'héritage néolithique*. Paris: Gallimard, 2011.

J.-D. Vigne. "Zooarchaeological aspects of the Neolithic diet transition in the Near East and Europe, and their putative relationships with the Neolithic demographic transition." T*he Neolithic Demographic Transition and Its Consequences*, J.-P. Bocquet-Appel and O. Bar-Yosef, eds. New York: Springer (2008): 179–205.

J.-N. Biraben. "*L'évolution du nombre des hommes*." *Population et Sociétés*, 394 (2003).

Acknowledgments

The idea for this book was born several years ago, during the writing of *Néandertal, mon frère* (*Neanderthal, My Brother*, Flammarion 2016), which provides a summary of our current understanding of the Neanderthal fossil population, its relationship with other fossil populations from the same period, and in particular with *Homo sapiens*.

It seemed necessary to us to show that our species did not appear suddenly. It is not the result of abrupt upheavals but, on the contrary, underwent a series of morphological changes that took place over a long period of time. Some of these changes, like bipedalism, occurred several million years ago. Above all, we wanted to show that culture had a strong influence on biological change. This is recognizable through activities such as production of stone tools, taming of fire, and social cooperation through oral communication, which guided our long process of becoming human.

This work has been the fruit of numerous of years of reflection during which we benefited from the support of different institutions and from the advice, aid, and encouragement of many friends and colleagues. For their help and advice in regard to the English language edition of this book, we would like to thank Professors Eric Delson and Ian Tattersall of the American Museum of Natural History, Professor Janet Monge of the University of Pennsylvania, and Professor Alan Mann of Princeton University.

We would like to thank Sophie Berlin and Christian Counillon, directors of scientific publications at Flammarion publishing house, whose encouragement was particularly important over the years we spent working on this book. We would also like to extend our thanks to Florence Giry of Flammarion's foreign rights department for her support and confidence. We are indebted to Matthew Lore, cofounder and proprietor of The Experiment, for publishing this book, and we sincerely thank Olivia Peluso for her assistance and patience, as well as Jennifer Hergenroeder for her help and Beth Bugler for the design, during the preparation of this book's American edition.

Index

NOTE: Page numbers in *italics* indicate a figure.

A

Acheulean culture, *63,* 66, 69
acoustic communication, 53
Africa as origin of humanity the-
    ories, 72–73. *See also* individual
    African countries and areas
agrarian lifestyle, 106–7, 115, 123,
    124, 125
Aiello, Leslie, 25, 34, 53
altruism, 79–80, 108–9
anatomical remodeling
    feet and toes, *4–5,* 7, 9, *11*
    hands, 42–43
    larynx and oral cavity/phonatory
        system, 55–56, 57
    for mobility vs. sedentary, 109
    for obligate upright locomotion,
        42–43
    pelvis, *4–5,* 7
    Sapien and chimpanzee com-
        pared, *4–5*
    stature, 17, 22–23, *24,* 25
animism, 104
antelope hunting technique, 48–49
Arabian peninsula, 85–86
Ardipithecus, 2, 6, 7, *8*
artists, subsidized, 103–4
Asia, *60–61,* 62. *See also* individual
    Asian countries and areas

A

Aurignacian culture, *63,* 103–5, 109
Australia, fossils in, 86–87
Australopithecus
    brain and gut size, 34
    footprints of, 9, *11,* 50
    fossils, 2, 7, 9, *10,* 12
    in hominin family tree, *8*
    Homo genus break from, 14,
        15–16
    lifestyle, 12
    proto-language feasibility, 56–57
    social grooming, 25
    tools potentially made by, vi,
        17–18, 19, *20,* 40
Australopithecus afarensis, 9, *10,*
    12–13, 15, 18, 21

B

basal metabolic rate, 32–33, 36
behavioral remodeling
    overview, 19, 21, 27, 73
    bonding techniques, 25–26
    cognitive revolution of *H. sapiens,*
        73–75
    decorated bodies, 104–5
    tribal enforcement of, 107–11
    See also culture; hunting

bipedalism
  overview, 1
  Australopithecus footprints, *9,
    11, 50*
  cycle of reinforcement, 23, 25
  evolutionary trigger for, 12–13
  hand/body evolution, 42–43
  and hominization, 21
  imperfect bipedalism, *3–5, 6–7*
  and language development, 51–54
  reasons for, 1, 14–15
  skull of arboreal vs. bipedal spe-
    cies, *4–5, 6*
  See also obligate bipedalism
Bocquet-Appel, Jean-Pierre, 123, 125
body temperature regulation system,
  43–44
bonobos, *3*
brain
  cerebral asymmetry and Broca's
    area, 56–57
  and cranial volume, 23, *24,* 25
  cycle of reinforcement, 23, 25
  eating animal proteins, 33–34
  energy demand, 32–33
  humans and chimpanzees com-
    pared, 28, *29,* 30
  of infants to 7-year-olds, 27–28
  prioritized over other organs,
    34, *35*
  ratio of brain to body weight, 23
  reconfiguration of, 30, 77
  and tool making, 40
  See also cognition
Bramble, Dennis, 48
Broca, Paul, 56
Broca's area, 56–57
Brunet, Michel, 6
Byrne, Richard, 52

C

Cain, Abel, Ötzi (Guilaine), 125, 127
calories burned, 14–15, 33
cannibalism, 116
Caspari, Rachel, 31
Catching Fire (Wrangham), 36–37
cerebellum, 42
Châtelperronian culture, *63*
Chauvet cave, France, 99, 103–4
cheaters, identifying, 81
childbirth, 28, 30, 125
Childe, Vere Gordon, 120
children
  overview, 27
  basal metabolic rate, 32–33
  brain growth first seven years,
    27–28, 30
  collaborative child rearing, 27,
    31, *35*
  and grandparents, 31
chimpanzees
  overview, 2, *3, 4–5*
  brain activity, 23, 28, *29,* 30, 33
  building nests in trees, 12
  empathy of, 81
  energy for locomotion, 14–15
  social grooming, 25
  tools of circumstance, 19
  treatment of cheaters, 81
China
  artifacts and fossils, 36, 62, 64–65
  Homo erectus in, *60–61*
  Sapiens in, 83, *88,* 89, 90–92
  warlords and warriors, 127
Clarkson, Chris, 86–87
climate change, 12–13
cognition
  chimpanzee capability, 23, 28, *29,*
    30, 33
  and cranial volume, 23, *24,* 25

of *H. sapiens,* 73
and hand movements, 41–42
as imitative capacity, 21–22
See also brain
communication
articulate language, 54–57
grooming as, 25, 51–52
hand signals and shouting, 47
involuntary emotions, 80–81
symbolic communication, 51–53
verbal communication, 53–54
with words, 53
Coppens, Yves, 12–13
cranial volume, 23, *24,* 25. *See also* brain
crossbreeding, *8,* 67, 74–75, 84, 92, 93–94, *95,* 96–97
cultural transmission, 21–22, 31, 78
culture
overview, 17, *63*
creating a "social glue," 102–5
cycle of modifications to biology and, 27
identifying via stone tool artifacts, 64–67
prehistoric material culture, 39
regional and tribal, 105
See also social structures; individual cultures
cycle of reinforcement, 23, 25

D
Darwin, Charles, 80, 108–9
Deloison, Yvette, 14, 50
Demeter, Fabrice, 90
Denisovans, *8, 45,* 66–67, 92, 93–94, *95, 96*
d'Errico, Francesco, 105
digestive system, 34
digital technology, 131
diseases, 123, *126*

Dmanisi, Georgia, fossils in, 64–65
domestication
animals for meat, 33–34
human beings, 107–11, 113–14
wolves, 105–7, 109–10
domestication syndrome, 106, 108–9
Dunbar, Robin, 25, 53
Dupont, Édouard-François, 116

E
East Africa-Asia-Australia Sapiens migration, 86–87, *88,* 89–91
ecological impact
human presence in nature, 100
hunting activities, 47–48, 75, *76, 77,* 97
overpopulation, 129
economic system, Neolithic, *121*
empathy, 78, 80–81, 108–9
encephalization coefficient, 23, 25
energy use, bipedal vs. arboreal, 14–15
England, fossils and artifacts, 116
epigenetics, 77
equality and distribution system, 111
Ethiopia, fossils in, *7, 18,* 21, 70
Europe
artifacts in, 36
Neanderthal genes in Europeans, 74–75
Ötzi (Iceman), 125, 127
European Neanderthals, 45, 66–67
expensive tissue hypothesis, 34, *35*
Expression of the Emotions in Man and Animals (Darwin), 80

F
fat enhancement in humans, 31–33
feet and toes, *4–5,* 7, 9, *11*
fight-or-flight response, 80–81
fire, mastering, 34, 36–37
Flores Man, 93

foods
  effect of cooking, 34, 36–37
  energy-rich, 33–34
  pemmican, 110
France, 99, 103–4, 105–6, 116
fur, loss of, 43–44

G
genetics
  of domesticated wolves, 105–7,
    109–10
  and language, 54–55
  Neanderthals, 74–75
  Sapiens, 74
Germany, cave art in, 104
globalization, beginning of,
  99–100
Gobekli Tepe, Turkey, 120
Goodall, Jane, 19
gorillas, 3, 23
grandparents' importance, 31
graves, 112–14, 117
Gravettian culture, 63, 105–6,
  109–14, 115–16, 122
great apes, 2, 3, 52
Great Rift Valley formation effect,
  12–13
grooming, 25, 51–52
Guilaine, Jean, 125, 127
Guy, Emmanuel, 103–4

H
hands, 9, 40–43, 42–43, 51–54
Harari, Yuval Noah, 74
Harmand, Sonia, 17–18
Harvati, Katerina, 84–85
Hayden, Brian, 111
Hobaiter, Catherine, 52
hominins and hominid family, 2,
  3–5, 6–7, 8, 13–14

hominization, vi, 17, 21–26. See also
  anatomical remodeling; behavioral remodeling; physiological
  remodeling
Homo genus
  African origin, 2, 70–72, 73
  brain prioritized over other
    organs, 34, 35
  break from Australopithecus, 14,
    15–16
  cerebral asymmetry and Broca's
    area, 56–57
  fire use, 36–37
  flat-faced man from Kenya, 18
  obligate bipedalism as defining
    characteristic, 13–14
  stature and cranial volume, 23,
    24, 25
  See also Sapiens
Homo erectus, 60–61, 62, 64, 69, 93
Homo ergaster
  overview, 47
  Asian form of, 62, 64–65
  brain and gut size, 34
  effective running/hunting, 50–51
  emotional reactions and empathy,
    81
  evolving into H. heidelbergensis,
    66, 69
  migration of, 60–61
  phonatory system, 55
  verbal communication, 53–54
Homo georgicus, 64
Homo habilis
  Asian form of, 64–65
  brain and gut size, 34
  emotional reactions and empathy,
    81
  knapping stone tools, 40
  larynx and oral cavity/phonatory
    system, 55–56, 57

shift to obligate bipedalism, 13–14

Homo heidelbergensis, 55, 66–67, 69–70, 73, 79–80

Homo neanderthalensis
emergence of, 66–67
light skin, 45
phonatory system, 55
social grooming, 25
See also Neanderthal species

Homo rudolfensis, 14

Homo sapiens
overview, 1, *8*
anthropological, cultural, and sociocultural traits, *101*
behavior in relation to nature, 75, *76,* 77, 100, 130
characteristics of, *101*
cultural evolution, 130–31
domestication of, 107–11, 113–14
emergence of, 67
empathy trait, 80–81
fossils, 71
infants' characteristics, 27–28, 77
migration from Africa, 73, 83, *88,* 91–92
in northern Eurasia, 92
obligate bipedalism, 13
value of socializing, 25–26
See also Sapiens

hordes or band societies, 69, 75, 78–82, 87, 92, 99, 100, 102, 107

Hublin, Jean-Jacques, 71

hunter-gatherer cultures, 33–34, 48–51, 78–81, 82, 102, 112, 125

Hunter-Gatherers, or the Origins of Inequality (Testart), 111

hunting
antelope, by San bushmen, 48–49
archaeological evidence of, 49–51
birds, with stones, 49

cooperation for, 51–57, 78
with dogs, 107
ecological impact, 47–48, 75, *76, 77,* 97
mammoths, 109–10
tribes formed for, 102

hunting camps, 110

hyoid bone, 55

I

India, fossils in, 90

India, Sapiens in, 89, 90

inequality
burial opulence, 112–14
and distribution system, 111
and war, 119, 127

infants, 125

infections and diseases, 123

isolation vs. the horde, 78

Israel
artifacts in, 36, 62, 64, 122–23
fossils in, 55, 70, 84

J

Jablonski, Nina, 44–45

Jebel Irhoud, North Africa, 70–71

Johanson, Donald, 12

Jouventin, Pierre, 107

K

Kappelman, John, 12

Keeley, Lawrence, 118

Kenya
first use of fire in, 36
fossils, 7, 17–18
Ileret site with *H. ergaster* footprints, 50
Koobi Fora site with butchering area, 50–51

Kenyanthropus platyops, 18

L

La Fontenelle, Jean de, 107
Lake Chad, Africa, fossils in, 6
language. *See* communication
Laos, fossil in, 90
Leakey, Mary, *11*
Lee, Sang-Hee, 31
Leroi-Gourhan, André, 104
Levallois culture, 70–71, 73
Levant corridor, 64, 72–73, 84–85,
    86, 91–92, *124*
Lewis, Margaret, 47–48
Lieberman, Dan, 48
Linnaeus, Carl, 1
Lomekwi 3 site, Kenya, 17–19, *20,*
    21, 39–40
Lomekwian stone culture, 18, 56, *63*
longevity, 31
Lucy *(A. afarensis),* 9, 12–13, 15, 18

M

Magdalenian tribe, *63,* 116–19
Man and Inequality (Hayden), 111
menopause, 31
metabolism, 32–33
migrations out of Africa
    H. sapiens, 73, 83, *88,* 91–92
    pre-Sapien, 59, *60–61,* 62
    Sapiens, 84–87, *88,* 89–93, 99
Miguelón *(Homo heidelbergensis),*
    79–80
millstones, 110
Misliya-1 site, Israel, 70, 73, 84
motor cortex, 41–42
Mousterian culture, *63*

N

Nadel, Dani, 122
natural resources
    H. sapiens exploitation, 75, *76,* 77,
        110–11

as impetus for bipedalism, 1, 14–15
    larger stature for exploiting,
        22–23, 25
    overexploitation of, 130
natural selection, tribal opposition
    to, 108–9
Neanderthal species
    ancestor of, 23, 54–55, 66–67
    cranial capacity, 23, *24*
    crossbreeding, 74–75, 84, 92,
        93–94, *95,* 96–97
    in Europe, 45, 66–67
    infants' characteristics, 77
    in Iraq and northern Eurasia,
        91–92
    Levallois technique, 70–71
    low population density, 92–93
    and Sapiens, in the Levant, 84
    similarity to *H. sapiens,* 74
    See also *Homo neanderthalensis*
neocortex, 28, *29,* 30
Neolithic period, *121,* 123, *124,* 125,
    126
Neolithic revolution theory, 120, 122
North Africa, fossils in, 70–72

O

obligate bipedalism
    overview, 39, 47
    and hominids, 13–14
    physiology changes, 27, 43–44
    See also bipedalism
obligate upright locomotion
    anatomical remodeling, 42–43
    loss of fur, 43–44
    running, 22, 43–44, 48, 50
Ohalo site, Israel, 122–22
Oldowan culture, *63*
orangutan, *3*
Orrorin tugenensis, 6–7, *8*
Ötzi (Iceman), 125, 127

P

Paranthropithecus, 8
Paranthropus, 2, *10*
pelvis, *4–5,* 7
phonatory system, 55–56
physiological remodeling
  overview, 27, 39
  adding animal proteins, 33–34
  basal metabolic rate, 32–33
  biological characteristics of *H. sapiens, 101*
  brain prioritized over other organs, 34, *35*
  childbirth, 28, 30, 125
  child's brain growth, 27–28, 30
  cranial volume, 23, *24,* 25
  enhanced fat, 31–33
  epigenetics, 77
  fur loss and sweat glands, 43–44
  infant development, 28
  longevity, 31
  skin, 43–45, 89
  sympathetic nervous system, 80
Pickford, Martin, 6–7
Pontzer, Herman, 33
population growth, 87, 89, 90, 119, 123, *126,* 129
predation. *See* hunting
prefrontal cortex, 30
prehistoric material culture, 39
prehistoric sciences discoveries, v–vi
Prehistory of Artistic Feeling, The (Guy), 103–4
Preneolithic period, 119–20, 122–23
"Prince of Sungir" tomb, Russia, 112–14, 115
proto-language, 56–57
Purgatorius (proto-primate), 2

R

Raichlen, David, 14
reproduction
  overview, 82
  childbirth, 28, 30, 125
  enhanced fat for, 31–32
  in hunter-gatherer culture, 125
  importance of, 79
  longevity aspects of, 30–31
  Sapiens hordes excelling at, 75, *76,* 77, 87
  tribal control of, 108
Roche, Hélène, 19
running upright, 22, 43–44, 48, 50. *See also* obligate upright locomotion
Russia, Gravettian tomb in, 112–14

S

Sapiens
  overview, 1–2, *3, 8,* 69, 73, 83, 129–31
  ancestor of, 23, 54–55, 66–67
  brain growth and plasticity, 77
  cerebral development, 28, *29,* 30
  crossbreeding, *8,* 67, 84, 92, 93–94, *95,* 96–97
  derivation of, 69–70
  domestic animals for meat, 33–34
  in Eurasia, 89, 96
  fossils, 69–72
  horde or band society, 78–82
  migration out of Africa, 84–87, *88,* 89–93, 99
  pre-Sapien migrations out of Africa, 59, *60–61,* 62
  See also *Homo genus; Homo sapiens*
Sapiens (Harari), 74
savannah hypothesis, 12–13
Scerri, Eleanor, 72

sedentary lifestyle, 32, 99, 109, 119–20, *121,* 122–24
self-awareness, 78–79
sense of touch, 41
Senut, Brigitte, 6–7
Shipman, Pat, 109
Siberia, fossils in, 92, 102–3, 106
skin, 43–45, 89
skull of arboreal vs. bipedal species, *4–5,* 6
slavery, 119
social structures
  and cerebral development in infants to children, 28
  collaborative child rearing, 27, 31, *35*
  hordes, 69, 75, 78–82, 87, 92, 99, 100, 102, 107
  increased cooperation of group members, 21–22, 25
  and size of neocortex, 25
  socializing around a fire, 37
  See also culture; individual species and groups
Solutrean culture, *63,* 116
South Africa, 9, 12, 48–49, 70
Svoboda, Jiří, 109
Swartkrans, Africa, 36
sweat glands, 43–44
symbolic communication, 51–53
sympathetic nervous system, 80
Syria, artifacts in, 120
Systema Naturae (Linnaeus), 1

T

Tanzania, *Australopithecus* footprints, 9, *11,* 50
taxonomy charts, 2, *8*
temples, 119–20
Testart, Alain, 111, 127
thumbs, 9

Tibet, fossils in, 67
tool making
  overview, 39
  by *Australopithecus* (likely), vi, 17–18, 19, *20*
  and brain size, 40
  as cultural phenomena, 18–19
  determining culture from, 64–67
  and evolution of cultures, *63*
  fire utilization, 37
  flint blades, 122
  Gravettian, 110
  and hominization, 22
  Levallois technique, 70–71, 73
  Lomekwian, 18, 56, *63*
  stone tool methods, *20*
  unique to humans theory, 39–40
totemism, 104
Toumaï *(Sahelanthropus tchadensis),* *4–5,* 6
tribes
  overview, 99–100, 102
  benefactors of war, 118–19
  decorated bodies, 104–5
  equality or inequality based on distribution system, 111
  human domestication, 107–11, 113–14
  infections and diseases, 123
  subsidized artists, 103–4
  temples, 119–20
  war and violence, 115–19
  wolf domestication, 105–7, 109–10
tropics, lack of fossils, 89–90
Turkey, artifacts in, 120

V

Vanhaeren, Marian, 105
Variation of Animals and Plants under Domestication (Darwin), 108–9

vitamin D absorption and skin color, 44–45
vonHoldt, Bridgett M., 105–6

W
Waal, Frans de, 81
war and the state
    overview, 115, 116, 118–19
    cold as priority, 116
    Gravettian culture, 115, 116
    Magdalenians, 116–19
    Neolithic period, 123
    warriors and kings, 127
War Before Civilization (Keeley), 118
Werdelin, Lars, 47–48
Wheeler, Peter, 34
Williams syndrome, 106
wolves, domestication of, 105–7, 109–10
women
    childbirth, 28, 30, 125
    enhanced fat, 31–32
    Gravettian culture, 112
    Magdalenian culture, 116
    males' tendency to protect, 79
    role assignments, 51, 82, 116
    as spoils of war, 119
    See also reproduction
Wrangham, Richard, 31–32, 36–37
Wu Liu, 90–91

Y
Yanomami tribe, the Amazon, 118

About the Authors

SILVANA CONDEMI, a paleoanthropologist, is the research director of the National Center for Scientific Research, the largest public scientific research organization in France, at Aix-Marseille University.

FRANÇOIS SAVATIER is a journalist for the magazine *Pour la Science* (the French edition of *Scientific American*), where he focuses on the science of the past.

Their previous book is the award-winning *Neanderthal, My Brother.*